T0291753

CAMBRIDGE LIBRARY COLLECTION

Books of enduring scholarly value

Astronomy

From ancient times, humans have tried to understand the workings of the world around them. The roots of modern physical science go back to the very earliest mechanical devices such as levers and rollers, the mixing of paints and dyes, and the importance of the heavenly bodies in early religious observance and navigation. The physical sciences as we know them today began to emerge as independent academic subjects during the early modern period, in the work of Newton and other 'natural philosophers', and numerous sub-disciplines developed during the centuries that followed. This part of the Cambridge Library Collection is devoted to landmark publications in this area which will be of interest to historians of science concerned with individual scientists, particular discoveries, and advances in scientific method, or with the establishment and development of scientific institutions around the world.

Greek Astronomy

From its beginnings in Babylonian and Egyptian theories, through its flowering into revolutionary ideas such as heliocentricity, astronomy proved a source of constant fascination for the philosophers of antiquity. In ancient Greece, the earliest written evidence of astronomical knowledge appeared in the poems of Homer and Hesiod. In the present work, first published in 1932, Sir Thomas Little Heath (1861–1940) collects some of the most notable essays and discussions of astronomical theory by Greek astronomers and mathematicians, presenting them in English translation for the modern reader. With chronological coverage, Heath's book features a thorough introduction, a doxography of what ancient authors said about the earliest theorists and longer excerpts exploring fundamental ideas. Among the pieces are extracts from Plato's *Republic* and Ptolemy's work on the impossibility of a moving Earth, alongside material from Aristotle, Euclid, Strabo, Plutarch and others.

Cambridge University Press has long been a pioneer in the reissuing of out-of-print titles from its own backlist, producing digital reprints of books that are still sought after by scholars and students but could not be reprinted economically using traditional technology. The Cambridge Library Collection extends this activity to a wider range of books which are still of importance to researchers and professionals, either for the source material they contain, or as landmarks in the history of their academic discipline.

Drawing from the world-renowned collections in the Cambridge University Library and other partner libraries, and guided by the advice of experts in each subject area, Cambridge University Press is using state-of-the-art scanning machines in its own Printing House to capture the content of each book selected for inclusion. The files are processed to give a consistently clear, crisp image, and the books finished to the high quality standard for which the Press is recognised around the world. The latest print-on-demand technology ensures that the books will remain available indefinitely, and that orders for single or multiple copies can quickly be supplied.

The Cambridge Library Collection brings back to life books of enduring scholarly value (including out-of-copyright works originally issued by other publishers) across a wide range of disciplines in the humanities and social sciences and in science and technology.

Greek Astronomy

T.L. Heath

CAMBRIDGE
UNIVERSITY PRESS

University Printing House, Cambridge, CB2 8BS, United Kingdom

Published in the United States of America by Cambridge University Press, New York

Cambridge University Press is part of the University of Cambridge.
It furthers the University's mission by disseminating knowledge in the pursuit of education, learning and research at the highest international levels of excellence.

www.cambridge.org
Information on this title: www.cambridge.org/9781108062800

This edition first published 1932
This digitally printed version 2014

ISBN 978-1-108-06280-0 Paperback

The Library of Greek Thought

GREEK ASTRONOMY

Edited by
ERNEST BARKER, Litt.D., D.Litt., LL.D.
PROFESSOR IN THE UNIVERSITY OF CAMBRIDGE

GREEK ASTRONOMY

BY

SIR THOMAS L. HEATH

K.C.B., K.C.V.O., F.R.S.,

*Sc.D. Camb., Hon. D.Sc. Oxford, Hon. Litt.D. Dublin,
Honorary Fellow (formerly Fellow) of Trinity College, Cambridge.*

LONDON & TORONTO

J. M. DENT & SONS LTD.

NEW YORK: E. P. DUTTON & CO. INC.

Printed in Great Britain
by The Temple Press Letchworth
for
J. M. Dent & Sons Ltd.
Aldine House, Bedford St. London
Toronto . Vancouver
Melbourne . Auckland
First Published 1932

PREFATORY NOTE

Most of the translations herein have been made by myself at different times. A large number of the passages were included in my sketch of the early history of Greek astronomy which forms the bulk of *Aristarchus of Samos, the Ancient Copernicus*, published in 1913 (Clarendon Press). It is my pleasant duty to offer my best thanks, first, to the Delegates of the Clarendon Press for allowing me to reproduce extracts from that work, and, secondly, to the same Delegates and the Provost of Oriel for giving me permission to make use of the Oxford translation of Aristotle. As I had myself translated all of the passages of Aristotle given in *Aristarchus of Samos*, as above mentioned, and also other passages in continuation of the same or in the same connexion, I have in these cases, for the sake of uniformity, used my own translations, so that I have not had occasion to do more than consult the Oxford translation.

I have also to thank Messrs. A. & C. Black Limited, for kindly allowing me to use some translations by the late Professor John Burnet (in his *Early Greek Philosophy*, fourth edition, 1930) of passages from the Doxography and other sources. I have appended Professor Burnet's name to the citations which I have made.

I desire to express my obligations to the *Geschichte der Sternkunde*, by Dr. Ernst Zinner of the Observatory at Bamberg. The publication of this highly important and comprehensive work in 1931 came in the nick of time for my purpose; I am particularly indebted to it for details of the ancient Egyptian and Babylonian astronomy.

T. L. H.

September 1932.

CONTENTS

INTRODUCTION

The first indications of astronomical knowledge on the part of the Greeks are to be found in the Homeric poems and the works of Hesiod. Homer mentions, in addition to the sun and moon, the Morning Star, the Evening Star, the Pleiades, the Hyades, Orion, the Great Bear ("which is also called by the name of the Wain and which turns round on the same spot and watches Orion; it alone is without lot in Oceanus' bath"), Sirius ("the star which rises in late summer . . . which is called among men 'Orion's dog'; bright it shines forth, yet is a baleful sign, for it brings to mortals much fiery heat"), the "late-setting Boötes" (Arcturus). With Homer the earth is a flat, circular disk; round it runs the River Oceanus, encircling the earth and flowing back into itself; from this all other waters take their rise, i.e. the waters of Oceanus pass through subterranean channels and appear again as the sources of other rivers. Over the flat earth is the vault of heaven, a sort of hemispherical dome exactly covering it. Below is Tartarus, covered by the earth and forming a sort of vault symmetrical with the heaven. It is not clear where the heavenly bodies go between their setting and rising; they cannot go under the earth because Tartarus is never lit up by the sun; possibly they are supposed to float round Oceanus, past the north, to the points where they next rise in the east.

There are vague uses of astronomical phenomena for the purpose of fixing localities or marking times of day and night, but there is little else in Homer that can be called astronomical.

Hesiod mentions practically the same stars as Homer, but he makes more use of celestial phenomena for the purpose of determining times and seasons in the year. With Hesiod the spring begins with the late rising of Arcturus (this would be in his time and climate the 24th February of the Julian calendar); sowing-time at the beginning of winter he fixed by the setting of the Pleiades in the early twilight or by the early setting of the Hyades or Orion (which would mean the 3rd, 7th, or 15th November according to the stars taken). Hesiod was acquainted with the solstices. Spring begins for him sixty days after the winter solstice; in the winter, he says, the sun turns towards the country of the dark-complexioned people, and rises later for the Greeks. He does not mention the equinoxes, but only remarks in one place that in late summer the days become shorter and the nights longer. He had an approximate notion of the moon's period; he puts it at thirty days.

So far as Greece is concerned, astronomy as a science, in however elementary a stage, begins with the Ionian philosophers, with whom Greek philosophy and Greek mathematics also begin. Ionia saw the first manifestations of the Greek genius for "inquiry," the insatiable curiosity of the race, their determination to know not only the fact but the why and wherefore of everything. As Professor John Burnet wrote: "No sooner did an Ionian philosopher learn half a dozen geometrical propositions and hear that the phenomena of the heavens recur in cycles than he set to work to look for law everywhere in nature and with an audacity amounting to ὕβρις to construct a system of the universe." Nor did the Greeks themselves hide their light under a bushel. "Let us," says the author of the Platonic *Epinomis*, "take it for granted that, whatever the Greeks take from the barbarians, they bring it to a finer perfection"; and Adrastus, a Peripatetic of the second century A.D., is quoted by Theon of Smyrna as saying, about the Chaldaeans and Egyptians in relation to astronomy in particular, that their methods were imperfect because they were destitute of *physiologia* (it is difficult to translate the word: the nearest equivalent would be the philosophy of nature and of natural *causes*), whereas *physical* considerations should enter into the inquiry: a condition which the Greek astronomers sought to fulfil, while taking from the barbarians certain facts which they had learnt from their observations of phenomena.

It is natural to ask how far the Greeks were indebted for the beginnings of their astronomy to the more ancient civilizations of Babylon and Egypt, not to speak of China. We can leave China out of account, notwithstanding the piquant story that, in 2159 B.C., the Chinese astronomers Hi and Ho were put to death *according to law* in consequence of an eclipse of the sun having occurred which they had not foretold.[1] It was the Babylonians and

[1] The sentence is quoted by Dr. Ernst Zinner: "The blind (musician) has drummed, the mandarins have mounted their horses, the people have flocked together. At this time Hi and Ho, like wooden figures, have seen nothing, heard nothing, and, by their neglect to calculate and observe the movements of the stars, have incurred the penalty of death."

Egyptians with whom the Ionian Greeks were in direct relations.

As regards Egypt and Babylon, we need not take too seriously the remarks of Simplicius in his commentary on Aristotle's *De caelo* that he had heard that the Egyptians possessed records of observations of the stars extending over 630,000 years, and that the Babylonians had them for 1,444,000 years, nor even the statement quoted by Simplicius from Porphyry that the observations sent from Babylon by Callisthenes at the request of Aristotle had been maintained for 31,000 years down to the time of Alexander. In any case it was long before the Babylonians had a satisfactory calendar, and, so long as this was the case, a trustworthy chronology was impossible. But a continuous record of dated observations began with the reign of Nabonassar, whose first year was our 747 B.C., "from which date," says Ptolemy, "we possess the ancient observations continued practically to the present day." A collection of observations, presumably including those from 747 B.C., was sent to Greece by Callisthenes, at the request of his uncle Aristotle, shortly after the capture of Babylon by Alexander in 331 B.C.; and these were without doubt accessible to Callippus. As regards the Babylonians, therefore, we have to take account not only of what the Ionian philosophers originally learnt from them, but of the more or less continuous effect on Greek astronomy of work which was being carried on simultaneously by Babylonian astronomers, some of whom, like Naburiannu (flourished perhaps about 500 B.C.) and Kidinnu (flourished 380, say), are worthy to be ranked with Hipparchus and Ptolemy.

Herodotus tells us that the Egyptians had a year of

365 days, and that they reckoned twelve months of thirty days each, but added five days in each year outside the total of 360 days. They divided the day into twelve hours and the night into twelve hours, so that the hours had different lengths according to the time of year. This affected the construction of the Egyptian water-clocks, which must have been in use from very early times for marking the hours of the night, since a definite improvement in them made about 1500 B.C. is on record. The Egyptians also had sun-clocks for use by day, on which different measurements of shadows were made to mark the hours. They knew a great number of stars or groups of stars, Sirius, Orion, the Great Bear, etc.; they began their year from the day after the early morning rising of Sirius. They knew of the zodiac circle, though whether they divided it longitudinally into parts is uncertain. At least forty-three constellations were known to them in the thirteenth century B.C. The five planets too were known in Egypt about the same date. It appears that, in the case of Mars, even his retrogradations were observed. Mars was called "the gleaming Horus," like the sun; Venus was at first called the planet of Osiris and afterwards the "Morning" and "Evening Star," which suggests that the identity of the two was recognized. Cicero says that the Egyptians called Venus and Mercury companions of the sun. Diodorus Siculus tells us that the priests of Thebes predicted eclipses quite as well as the Chaldaeans; it is possible that, like the Greeks, the Egyptians learnt from the Chaldaeans the method on which they proceeded.

The Greeks owed much more to the Babylonians than to the Egyptians; now that the matter has been more studied (as it has been in the last few years) from the Babylonian side, the debt is seen to be much greater than was formerly supposed.

As early as the second millennium B.C. the Babylonians recognized the zodiac as the circle in which the planets move. They divided it into signs of 30 degrees each; and, when we are told by Pliny that Cleostratus of Tenedos, who belongs probably to the second half of the sixth century B.C., "recognized the signs in it," it is a fair inference that Cleostratus imported from Babylon into Greece the knowledge of the zodiac and the constellations in it, and perhaps of some other constellations.

In the earliest times the Babylonians seem to have had a rough year of 360 days (twelve months of thirty days each), but it was long before they had a regular system for intercalating months; this was done, whenever it seemed desirable, on the instructions of some recognized authority. The oldest lunar cycle known is that of 223 lunar months or 6585⅓ days, which Suidas calls the *saros*; this is the period after which eclipses recur. Now from 747 B.C., when the dated record of observations began, to the date of the eclipse foretold by Thales (585 B.C.) there was a period of 162 years, and there can be no doubt that, by the time of Thales, the Babylonians had discovered the period in question, and that this period had become known to Thales and was the basis of his prediction.

Cleostratus, again, is said to have been the first to construct the *octaëteris*, or eight-year cycle. This was apparently the first attempt to discover a period which would satisfy the three requirements of containing at once an exact number of days, of lunar months, and of solar years. Eight solar years with $365\frac{1}{4}$ days to the year gave 2922 days, and 99 months. This cycle was in use at Babylon from 528 to 505 B.C., which would be in the time of Cleostratus; it is probable therefore that he imported the idea thence. From 505 B.C. the Babylonians used an intercalation-period of twenty-seven years including ten intercalary months; but from 383 B.C. they used the better cycle of nineteen years including seven intercalary months. According to Diodorus, Meton put forward his nineteen-years period at Athens in the year 433/2 B.C.; this is doubtful, seeing that the first attested use of the period in Greece was in 342 B.C.

About 1000 B.C. the Babylonians determined the time of day by means of a sundial consisting of an upright pointer one ell in length standing on a plane base; this the Greeks called a *gnomon*. Another form of dial had a pointer fixed upright in the middle of a hemispherical bowl; the Greek name of this was *polos*. According to Herodotus the Greeks obtained their knowledge of both these instruments and of "the twelve parts of the day" from the Babylonians.

The Babylonians expressed the angular distances between two stars in terms of *ells* and *dactyli* (fingers); an ell is the equivalent of two degrees, and there were twenty-four *dactyli* to the ell. This method of measuring angles was used by Hipparchus; and of course the sexagesimal system of degrees, minutes, seconds, etc., used by Ptolemy was Babylonian in origin.

Observations of the planets by the Babylonians go back to the second millennium B.C. Apparently Venus was the first to be studied; there are Venus-tables based on observations made between 1921 and 1901 B.C. Jupiter, too, and Mars were observed. The Babylonians realized that the movements of the planets showed irregularities as compared with those of the sun and moon; but they do not seem to have discovered the stationary points and retrogradations. The Babylonians described the colours of stars in terms of the colours of the several planets, using first four of them for this purpose, then five including Venus. Where a star was not exactly of the colour of one of the planets, they used more than one of the planetary colours to describe it. Some part at least of what the Babylonians had learnt about the planets would doubtless become known to the Ionian Greeks.

THALES, the first of the Ionian philosophers, was a native of Miletus and lived probably from about 624 to 547 B.C. Statesman, engineer, mathematician, astronomer, and man of business, he was distinguished in almost every field of human activity. He was one of the Seven Wise Men. Many sayings are attributed to him. "Know thyself" is one of them. Another is: "Of all things that are, the most ancient is God, for he is uncreated; the most beautiful is the universe, for it is God's workmanship; the greatest is Space, for it contains everything; the swiftest is Mind, for it speeds everywhere; the strongest is Necessity, for it masters all; and the wisest Time, for it brings everything to light."

Thales' claim to a place in the history of astronomy rests almost entirely on one achievement attributed to him, that of predicting an eclipse of the sun which took place during a battle between the Lydians and the Medes and was probably that of 28th May 585 B.C. It seems clear that this was a prediction of the same sort as the Chaldaeans were able to make, and that it was based on knowledge, obtained by Thales from the Babylonians, of the period of 223 lunations after which eclipses recur. The method served very well for lunar eclipses, but it would very often fail for solar eclipses because it took no account of parallax. It was therefore great luck for Thales that his eclipse should have been visible and total at the scene of the battle in question.

The following further details of Thales' astronomical knowledge are handed down. He gave the length of the year as 365 days; this he probably learnt from Egypt. He is said to have discovered the inequality of the four astronomical seasons, that is, the four parts of the tropical year as divided by the solstices and equinoxes, and to have written on the Solstice and on the Equinox; here too he may have been indebted to the Egyptians. He is said to have observed the Little Bear and to have used it for finding the pole; he advised the Greeks, so we are told, to follow the practice of Phoenician navigators, who steered their course by the Little Bear, whereas the Greeks sailed by the Great Bear. A handbook under the title of *Nautical Astronomy* was alternatively attributed to Thales and Phocus of Samos. This was no doubt intended to be an improvement on the *Astronomy* attributed to Hesiod, and was followed in its turn by the *Astrology* of Cleostratus, which probably dealt with the risings and settings of various stars and groups of stars.

Thales' interest in astronomy is attested by the well-known story that he fell into a well while star-gazing, and was rallied (as Plato has it) by a clever and pretty maidservant from Thrace for "being so eager to know what goes on in the heavens that he could not see what was in front of him, nay at his very feet."

Thales' theory of the universe was this. According to him, the one "first principle" (as Aristotle calls it) or material cause of all things is water; earth is the result of condensation of water, air is produced from water by rarefaction, and air again when heated becomes fire. We may assume therefore that, in Thales' view, there was in the beginning only the primordial mass of water and from this other things were gradually differentiated. Thales said that the earth floats on the water like a log or a cork; he would therefore presumably regard it as a flat disk or a short cylinder. Simplicius, the commentator on Aristotle, conjectures that Thales derived his ideas from myths current in Egypt. Paul Tannery pointed out the similarity between Thales' view of the world and that contained in ancient Egyptian papyri. According to these, there existed in the beginning the *Nu*, a primordial liquid mass in the limitless depths of which floated the germs of things. When the sun began to shine, the earth was flattened out, and the waters separated into two masses. The one gave rise to the rivers and the ocean; the other, suspended above, formed the vault of heaven, *the waters above*, on which the stars and the gods, borne by an eternal current, began to float. The sun, standing upright in his sacred bark which had endured for millions of years, glides slowly, conducted by an army of secondary gods, the planets, and the stars. We may compare also the first chapter of Genesis, verses 6 to 10: "And God said, Let there be a firmament in the midst of the waters, and let it divide the waters from the waters. And God made the firmament, and divided the waters which were under the firmament from the waters which were above the firmament; and it was so. And God called the firmament Heaven. And God said, Let the waters under the heaven be gathered together unto one place, and let the dry land appear; and it was so. And God called the dry land Earth; and the gathering together of the waters called he Seas." The Babylonian account of creation contains apparently the same idea of the primordial watery chaos being cleft into two parts, the chaos, however, being personified as a monster which Marduk, the supreme God of Babylon, cleaves in twain with his scimitar. So far, therefore, as his views on the universe are concerned, Thales was not greatly in advance of the Egyptians, the Hebrews, and the Babylonians.

The case is quite different with ANAXIMANDER (about 611–546 B.C.), an associate of Thales. Here we find the Greek speculative genius in full career. We are told that Anaximander wrote only one book, which was *On Nature*, and that he postponed its appearance until he was sixty-four years of age. This seems to indicate a determination on his part that the work should represent his maturest thought on the subject.

Almost all that we know of Anaximander's doctrines comes ultimately from the great work of Theophrastus entitled *Physical Opinions*. Anaximander's first principle or material cause was not water nor any other of the so-called elements, but what he called the Infinite or Unlimited, simply. The nature of this primordial substance he did not define; it was a kind of indeterminate stuff out of which "the heavens and all the worlds," and all that is in them, could be generated. Out of it all the worlds arise and into it they must all pass away once more.

It was necessarily infinite "in order that the process of coming into being might not suffer any check." At any one time there exists an infinite number of worlds, in different stages respectively; some are coming into being, some are at their prime, some are passing away; "for existent things must pay the penalty and make reparation to one another for the injustice they have committed, according to the sequence of time" — as if, by being separated out and becoming independent, things took an unfair advantage! In addition to his Infinite substance, Anaximander postulated eternal motion, because without motion there can be no coming into being or passing away.

In the case of our world, the portion of the Infinite which was separated off to form it was first separated into two opposites, the Hot and the Cold. The hot appeared as a sphere of flame which "grew round the air about the earth as the bark round a tree." Then the sphere was "torn off and became enclosed in certain circles or rings, and thus were formed the sun, the moon, and the stars." These rings were a sort of circular hoops or tubes made of compressed (and opaque) "air" enclosing fire within them throughout their length and allowing the fire to be seen at one place only, where there is a circular vent through which the fire shines out, so producing the appearance of the heavenly body. Sun, moon, and stars are therefore like gas-jets, as it were.

There is some doubt as to the nature of the eternal motion postulated along with the Infinite or Unlimited. The separating out of opposites suggests some process like shaking and sifting as it were in a sieve. On the other hand, when we are told that a spherical shell of flame grew all round the air about the earth, this seems to imply that it was rotatory. Again, there is no positive evidence that Anaximander's notion was that of a vortex such as was assumed by Anaxagoras and others later; but this must have been the way in which the earth came together in the centre, heavy things tending to the centre in a vortex or eddy.

The circles or hoops containing respectively one vent through which the fire shines out must apparently be conceived as revolving like wheels about some axis passing through the centre of the earth. No doubt those carrying the fixed stars would be parallel and rotate about a common axis; but Anaximander says that the wheels of the sun and moon lie obliquely. This shows that Anaximander was aware of the obliquity of the ecliptic; and the wheels of the sun and moon must be supposed to rotate about an axis perpendicular to the plane of the ecliptic. Eclipses of the sun and moon occur through the openings by which the fire finds vent being stopped up; "the moon appears sometimes as waxing, sometimes as waning, to an extent corresponding to the opening or closing of the passages."

Anaximander boldly declared that the earth is suspended freely, without support; it remains in its position, he said, because it is at an equal distance from all the rest of the heavenly bodies. By this he clearly meant that the earth is in equilibrium. Aristotle took the principle involved to be that of "indifference," and has an amusing criticism which will be quoted in its proper place. According to Anaximander, the earth is a short cylinder, one of its two plane faces being that on which we stand, and the other opposite to it; its depth is one-third of its breadth.

Furthermore, Anaximander was the first to speculate about the sizes and distances of the sun and moon. He said that the moon's ring is nineteen times the size of the earth, and the sun's ring twenty-eight (or twenty-seven) times the size of the earth; hence we may suppose that he took the moon's distance to be nineteen (or eighteen) times the radius of the plane face of the earth, and the sun's distance twenty-eight (or twenty-seven) times the same radius. The sun itself he held to be equal in size to the earth.

Anaximander is said to have set up in Sparta a *gnomon* (a sundial with a vertical needle on a plane base) and to have marked on it the solstices, the times, the seasons, and the equinox. As we have seen, Herodotus says that the Greeks learnt from the Babylonians about the *polos* and the *gnomon* and the twelve parts of the day.

Anaximander has yet another claim to undying fame. He was the first to draw a map of the inhabited earth. The Egyptians had drawn maps before, but only of particular districts; Anaximander boldly planned out the whole world "with the circumference of the earth and of the sea." Corrected by Hecataeus, this map became the object of general admiration.

Most remarkable of all perhaps (though irrelevant to our subject) is the fact that Anaximander hinted at a theory of evolution. According to him, animals first arose from slime evaporated by the sun; they first lived in the sea and had prickly coverings; men at first resembled fishes (or were born *inside* fishes).

Anaximenes of Miletus, an "associate" of Anaximander, lived from about 585/4 to 528/4 B.C. He made "air" the primary element from which all things are evolved. He held that the earth was flat, like a table, and supported by the air. According to him, the sun, moon, and stars were originally evolved from earth; it is from earth that moisture arises; then, when this is rarefied, fire is produced; the stars are composed of fire that has risen aloft. The sun, moon, and stars are all made of fire and ride on the air because of their breadth; the sun is flat like a leaf. We are told that Anaximenes also held that the stars are fastened to a crystal sphere, like nails or studs; the stars therefore which ride on the air must in that case be the planets. The stars, however, he said, do not move or revolve under the earth, as some suppose, but pass laterally round the earth, just as a cap can be turned round the head. He does not explain how this can be if the stars are fixed on the sphere of the heaven like nails; and it may be that, with Professor John Burnet, we should reject the testimony of Aëtius to the latter effect.

Anaximenes made one innovation of significance; he held that there are also, in the region occupied by the stars, bodies of an earthy nature which are carried round with them. If, as seems likely, these bodies were separate from the stars, they were probably invented to explain eclipses of the sun and moon.

"Just as our soul, being air, holds us together, so do breath and air encompass the whole world." In this idea Anaximenes anticipated Pythagoras.

PYTHAGORAS, born at Samos about 572 B.C., was probably the first to hold that the earth is spherical in shape. He is also said to have been the first to declare that the planets have a movement of their own independent of, and in a sense opposite to, that of the daily rotation;

Alcmaeon, a physician of Croton, was also credited with a like statement. It seems certain, however, that the fact in question was learnt from Babylon or Egypt. The same is probably true of the "discovery", alternatively attributed to Pythagoras and Parmenides, that the Morning Star and the Evening Star are one and the same.

Pythagoras seems to have held that the universe, like the earth, is spherical in shape. He would most likely give the same shape to the sun, moon, and stars. There is no reason to doubt that, for Pythagoras himself (as distinguished from the later Pythagoreans), the earth was still at rest at the centre of the universe, while the sphere of the fixed stars had a daily rotation about its axis from east to west.

In comparison with the views of the Ionian philosophers and Pythagoras, the ideas of Xenophanes of Colophon and Heraclitus of Ephesus seem very primitive and crude.

XENOPHANES was born about 570 and died after 478 B.C. He spoke of Pythagoras in the past tense, and he says of himself that, from the time when he was twenty-five years of age, three-score years and seven had "tossed his careworn soul up and down the land of Hellas."

Xenophanes, too, had views about the origin of our system. The world, he said, was evolved from a mixture of earth and water, and the earth will gradually be dissolved again by moisture. This he inferred from the fact that shells are found far inland and on mountains, and in the quarries of Syracuse there have been found imprints (i.e. fossils) of fishes and seaweed, and so on, these imprints showing that everything was covered in mud long ago, and that the imprints dried on the mud.

All men will disappear when the earth is absorbed into the sea and becomes mud, after which the process of coming into being must start again; all the worlds alike suffer this change. This reminds us of Anaximander.

For Xenophanes the sun, moon, and stars are clouds set on fire; clouds formed from moist exhalations take fire, and the sun is formed from the resulting fiery particles collected together; so with the moon, except that the cloud is here described as "compressed"; the moon's light is its own. When the sun sets, it is extinguished, and, when it next rises, it is a new one; it is likewise extinguished when there is an eclipse. There are many suns and moons according to the regions, divisions, and zones of the earth; and at certain times the (sun's) disk comes upon some uninhabited division of the earth, and so, as it were stepping on vacancy (or void), suffers eclipse. The sun goes forward *ad infinitum*; it only appears to revolve owing to its distance from us. The earth is flat; on its upper side it touches the air; on the under side it extends without limit. The earth is infinite in extent, and is surrounded neither by the air nor by the heaven.

Heraclitus of Ephesus (born about 544/0 B.C.) took fire to be the original element. Fire condenses into water, and water into earth; this is the downward course. The earth, on the other hand, may partly melt; this produces water, and water vaporizes into air and fire; this is the upward course. "All things are in flux"; you cannot step twice into the same stream.

There are two kinds of exhalations, said Heraclitus, arising from the earth and from the sea; one kind is bright and pure, the other dark; night and day, the months, seasons, years, the rains, the winds, etc., are all produced by the variations in the proportion between the two exhalations. In the heavens are certain basins or bowls turned with their concave sides towards us, which collect the bright exhalations; these are the stars. The sun and the moon are bowl-shaped like the stars, and similarly lit up. The flame of the sun is the brightest and hottest, the other stars are farther away and therefore give less light and warmth. "If there were no sun, it would be night for anything the stars could do." The moon, being nearer the earth, moves in less pure air, and is consequently dimmer than the sun. The sun and the moon are both eclipsed when the bowls are turned upwards (i.e. with the concavities away from us); the changes in the moon are caused by gradual turnings of the bowl. There is a new sun every day.

Heraclitus held (as did Epicurus afterwards) that the diameter of the sun is one foot, and that its actual size is the same as its apparent size. It would seem that he said nothing about the earth.

Parmenides of Elea was born in the second half of the sixth century B.C., later rather than earlier. Certain things are attributed alternatively to Pythagoras and Parmenides, such as the discovery of the spherical shape of the earth, the division of the earth into zones, and the recognition of the identity of the Morning and Evening Stars. Tradition says that Parmenides had been a Pythagorean; it is therefore not surprising that his cosmology was on Pythagorean lines, with some differences.

It seems certain that Pythagoras himself conceived the universe to be a sphere and attributed to it daily rotation about an axis; this involved the assumption that it is itself finite, but that something exists outside it; the Pythagoreans were therefore bound to hold that, beyond the finite rotating sphere, there is limitless void or empty space; this agrees with their notion that the universe in a sense *breathes*, a supposition which may be attributable to the Master himself. Parmenides, on the other hand, denied the existence of the infinite void, and was therefore obliged to make his finite sphere representing Being or truth (though he does not use the word Being) motionless, and to maintain that its apparent rotation is an illusion. Like Pythagoras, he declared the earth to be spherical, and he is associated with Democritus as having argued (as did Anaximander) that the earth remains in the centre because, being equidistant from all points on the sphere representing the universe, it is in equilibrium, and there is no reason why it should tend to move in one direction rather than another.

The physical theories of Parmenides followed, in the main, one or other of the Ionian philosophers. The earth, he said, was formed from a precipitate of condensed air. The stars he regarded as "compressed" fire. The sun, moon, planets, and other stars he arranged in a system of what he called "wreaths" or bands round the earth, which are like the hoops of Anaximander in some respects but not in others. Anaximander had distinguished separate hoops belonging to the sun, moon, and stars respectively, the hoops being of different sizes, the sun's the largest, the moon's the next, and those of the stars smaller still. Parmenides' "wreaths" were one above the other relatively to the earth as centre. That which encloses the All is solid like a wall; this is the solid firmament. Below this is a band of fire, which is the aether-fire. In the very middle of all is a solid thing which is the earth's crust, and next, encircling it, there is a wreath of fire, the inner side of which is apparently our atmosphere. Between the outer and inner wreaths of fire are mixed wreaths or bands consisting of light and darkness in combination. These exhibit the phenomenon of fire shining out here and there, and they include the Milky Way, as well as the sun, moon, and planets. The constitution of the heavenly bodies Parmenides explains in this way: "The air is thrown off the earth in the form of vapour owing to the violent pressure of its condensation; the sun and the Milky Way are an exspiration of the fire; the moon is a mixture of both elements, air and fire."

Aëtius says that Parmenides declared the moon to be illuminated by the sun. This statement may have been based on two lines from Parmenides' poem, which have commonly been thought (e.g. by Burnet) to support it. The first of the lines speaks of the moon as "a night-shining foreign light wandering round the earth."[1] The other line describes the moon as "always fixing its gaze on the beams of the sun." The word for "foreign" in the first line is translated "borrowed" by Burnet, but it need not necessarily mean "borrowed." The second line is a statement of a fact but scarcely implies knowledge of the cause. Further, Plato in the *Cratylus* speaks of "the fact which he (Anaxagoras) recently asserted, namely that the moon has its light from the sun." It may be, as Burnet says, that this was only the first intro-

[1] The expression "foreign light" is a witty adaptation of Homer's ἀλλότριος φῶς, which is used, not of a light, but of a man, and simply means "stranger."

duction of the idea into Athens; but on the whole it seems better to assign the discovery to one (Anaxagoras) who said that the moon "is of earthy nature and has in it plains and ravines," than to one (Parmenides) who made it of fire, or of air and fire.

Speaking of Empedocles and Anaxagoras, Aristotle says that Anaxagoras was before Empedocles in age but "after him in works." This may mean simply that Anaxagoras wrote later than Empedocles; but Theophrastus apparently took the phrase to mean "inferior in achievements." If the latter was the meaning, we must put it down to prejudice on the part of Aristotle.

Empedocles was a great democratic leader at Acragas (Agrigentum), being apparently the only native citizen of a Dorian state who plays an important part in the history of philosophy. All that we know of his date is that he was born a little before 490 B.C. and died after 444 B.C. at the age of sixty.

Empedocles held that there are four original elements, and that everything comes about through their combination and separation respectively by two moving principles, Love and Strife, acting alternately. His ideas in astronomy were crude. According to him, the heaven is a crystal sphere, and the fixed stars are attached to it, while the planets are free. The sphere, which is "solid and made of air condensed or congealed by the action of fire like crystal," is, however, not quite spherical, the height from the earth to the heaven being less than its distance from it laterally, and the universe being therefore shaped "like an egg" (it should apparently rather have been like an oblate spheroid—the solid described by the revolution of an ellipse about its minor, not its major, axis). The sun's course is round the extreme circumference of the world; it is prevented from moving always in a straight line by the resistance of the enveloping sphere and by the tropic circles.

Empedocles thought that the earth is kept in its place by the swiftness of the revolution of the heaven, just as we may swing a cup with water in it round and round so quickly that in some positions the cup may actually be turned downwards without the water being spilt.

Empedocles explained the succession of day and night in this way: Within the crystal sphere, and filling it, is a sphere consisting of two hemispheres, one of which is wholly of fire and therefore light, while the other is a mixture of air with a little fire, which mixture is darkness or night. The revolution of these two hemispheres round the earth produces at each point on its surface the succession of day and night. The beginning of this motion was, he said, due to the pressure of the mass of fire in one of the hemispheres upsetting the equilibrium of the heaven and causing it to revolve.

Perhaps even more strange is his account of the sun. The sun is, in its nature, not fire, but a reflection of fire. His idea was apparently that the mass of the fire in the fiery hemisphere is reflected from the earth upon the crystal vault, the reflected rays being concentrated in what we see as the sun. The sun, which consists of the said reflection, is, he says, equal in size to the earth.

Yet Empedocles deserves a place in the history of astronomy and science on account of one thing—his theory that light travels and takes time to pass from one point to another. For this we have the authority of Aristotle: "Empedocles represented light as moving in space and arriving at a given point of time between the earth and that which surrounds it, without our perceiving its motion." But he had no better argument to oppose to Empedocles than that, "though a movement of light might elude our observation within a short distance, that it should do so all the way from east to west is too much to assume."

ANAXAGORAS, born about 500 B.C. at Clazomenae, in the neighbourhood of Smyrna, neglected his possessions, which were considerable, in order to devote himself to science. Asked by someone what was the object of being born, he replied: "To investigate the sun, moon, and heaven." He seems to have been the first philosopher to take up his abode at Athens; there he enjoyed the friendship of Pericles. When Pericles became unpopular shortly before the outbreak of the Peloponnesian War, he was attacked through his friends. Anaxagoras was accused of impiety for saying that the sun was a red-hot stone and the moon earth. Some say that he was fined and banished, others that it was intended to put him to death and that he was with difficulty saved by Pericles. Ultimately he went and lived at Lampsacus, where he died at the age of seventy-two.

One epoch-making discovery is attributed to him, namely that the moon does not shine by its own light but receives its light from the sun. Thus he was able correctly to account for eclipses of the sun as due to the interposition of the moon, and for eclipses of the moon as due to the interposition of the earth. He thought, however (like Anaximenes), that there were also other dark bodies invisible to us which sometimes obscured the moon and caused eclipses.

Anaxagoras' cosmogony was full of fruitful ideas in which we recognize remarkable resemblances to the root-assumptions in the nebular hypothesis on which the latest cosmological theories are based.

According to Anaxagoras, the formation of the world began with a vortex set up, in a portion of the mixed mass in which (as he said) "all things were together," by his one motive principle, *Nous* or Mind. This rotatory movement began at one point and then gradually spread, taking in wider and wider circles. The first effect was to separate two great masses, one consisting of the rare, hot, light, dry, called the "aether," and the other of the opposite categories and called "air." The aether or fire took the outer position, the air the inner. The next step is the successive separating, out of the "air," of clouds, water, earth, and stones. The dense, the moist, the dark and cold, and all the heaviest things collect in the centre as the result of the circular motion; and it is from these elements when consolidated that the earth is formed. But, after this, "in consequence of the violence of the whirling motion, the surrounding fiery aether tore stones away from the earth, and kindled them into stars." Reading this with the remark that "stones rush outwards more than water," we see that Anaxagoras conceived the notion of a *centrifugal* force as distinct from that of concentration brought about by a motion like that of an eddy, and further that he assumed a series of projections or hurlings-off of precisely the same kind as the theory of Kant and Laplace assumes for the formation of the solar system. We may assume that Anaxagoras' hypothesis was not mathematically grounded as was that of Laplace. Under the latter hypothesis (to use Sir James

Jeans' words) "a hot nebulous mass of gas is supposed to be flattened in shape continually as it cools. In accordance with the principle of conservation of angular momentum, the diminution in size must be accompanied by a continual increase in velocity of rotation. When the rotation has reached a certain speed, centrifugal force will, according to Laplace, outweigh the gravitational attraction of the mass at its equator, and the result is supposed to be that a ring of matter is thrown off from the equator." There is, naturally, in Anaxagoras nothing about conservation of angular momentum, but there is in the assumption of stones being torn off by the violence of the whirling motion a vague suggestion of centrifugal force overpowering a tendency on the part of the heaviest parts of the whirling mass to collect in the centre.

Apart from his cosmogony, Anaxagoras did not make much advance upon the crude Ionian astronomy. He thought the earth was flat and supported by the air. According to him, the sun, the moon, and all the stars are stones on fire, which are carried round by the revolution of the aether. The sun "is larger than the Peloponnese." The moon is of earthy nature, and has in it plains, mountains, and ravines. Both Anaxagoras and Empedocles thought that the axis of the world was originally perpendicular to the surface of the (flat) earth, the visible north pole being at the zenith, and that it was displaced afterwards.

One of the fragments of Anaxagoras' writings leaves no room for doubt that he conceived that there were other worlds than ours. "Men were formed," he says, "and the other creatures which have life; the men too have inhabited cities and cultivated fields as with us;

they have also a sun and a moon and the rest, as with us, and their earth produces for them many things of various kinds, the best of which they gather together into their dwellings and live upon. Thus much," he adds, "I have said about separating-off, to show that it will not only be with us that things are separated off but elsewhere as well."

A new and startling innovation in theoretical astronomy is due to the successors of Pythagoras in the Pythagorean school. This was nothing less than the abandonment of the geocentric hypothesis and the reduction of the earth to the status of a planet like the others. Aëtius (probably on the authority of Theophrastus) attributes the resulting system to Philolaus (second half of fifth century B.C.), Diogenes Laërtius to one Hicetas of Syracuse, Aristotle to "the Pythagoreans" simply. The system falls short of the Copernican in that the earth and the planets do not revolve about the sun, but, with the sun and moon, revolve round an assumed central fire invisible to us. The scheme is as follows. The universe is spherical in shape and finite in size. Outside it is infinite void. At the centre is the central fire, the Hearth of the Universe, called by various names, such as the Tower or Watch-tower of Zeus, the Throne of Zeus, the Mother of the Gods, and here is located the governing principle, the force which directs the movement and activity of the universe. In the universe there revolve about the central fire the following bodies, ten in number. Nearest to the central fire revolves a body called the "counter-earth" which always accompanies the earth, the orbit of the earth coming next to that of the counter-earth; next to the earth, reckoning in order from the centre outwards, comes the moon, next to the moon the sun, next to the sun the five planets, and, last of all, outside the orbits of the planets, the sphere of the fixed stars. The counter-earth, which accompanies the earth and revolves in a small orbit, is not seen by us because the hemisphere on which we live is turned away from the counter-earth, and therefore from the centre. Incidentally this involves a rotation of the earth about its axis completed in the same time as it takes the earth to complete a revolution about the central fire. What was the object of introducing the counter-earth? Aristotle supplies the probable answer when he says that eclipses of the moon were considered to be due to the interposition, sometimes of the earth, sometimes of the counter-earth (not to mention the other dark bodies assumed by Anaximenes and Anaxagoras); the counter-earth was probably invented to account for the frequency of lunar eclipses as compared with solar.

The earth is said, by its revolution round the central fire, to produce day and night, and it is a fair inference that the earth is supposed to complete one revolution about the central fire in about twenty-four hours. In that case, from the point of view of producing day and night, the revolution of the sphere of the fixed stars is unnecessary. Possibly this point had not occurred to the authors of the system.

Oenopides of Chios was a younger contemporary of Anaxagoras. He was a geometer of note as well as an astronomer. Eudemus' authority is quoted by Theon of Smyrna for the statement that Oenopides discovered the obliquity of the ecliptic; but the idea at all events was, as we have seen, present to Anaximander. The estimate of twenty-four degrees for the inclination of the ecliptic was made before Euclid's time but was apparently not given by Oenopides, since, according to Theon, it was "other astronomers" who found that the measure of the obliquity was the angle subtended at the centre of a circle by the side of a regular fifteen-sided figure inscribed in it (cf. Eucl. IV, 16).

We come next to the atomists Leucippus and Democritus. Leucippus was a contemporary of Anaxagoras and Empedocles. Democritus of Abdera was, as he himself says, "young when Anaxagoras was old"; he was born in 470/69 or 460/59 B.C., and lived to a great age (90, 100, 104, 108, or 109, according to different accounts).

The cosmological views of LEUCIPPUS are described by Theophrastus as follows. The worlds, unlimited in number, arise through "bodies," that is, atoms, falling into the void and meeting one another. By abscission from the infinite many "bodies" of all sorts of shapes are borne into a great void, and their coming together sets up a vortex. By the usual process, in the case of our world, the earth collects at the centre. The lighter atoms pass into the empty space farther out as if they were being winnowed. The outer shell becomes thinner and thinner, but, on the other hand, it is made larger by the influx of atoms from the outside, adding to itself whatever atoms it touches. Of these some are locked together and form a mass, at first damp and miry, but, when they have dried and revolve with the universal vortex, they then take fire and form the substance of the stars.

The earth is like a tambourine in shape and rides or floats by reason of its being whirled round in the centre.

The sun revolves in a circle; the circle of the sun is the outermost, that of the moon is nearest to the earth, and the circles of the stars are between.

The ideas of Leucippus on astronomy thus show no advance on those of his predecessors.

Most of the views of DEMOCRITUS are a restatement of those of Anaxagoras. As regards the order of the sun, moon, and planets in space, we may notice that, whereas Anaxagoras made the sun come next to the moon in proximity to the earth, Democritus made the order, reckoning from the earth, to be: moon, Venus, sun, the other planets, the fixed stars.

PLATO's views on astronomy have to be gathered from several different dialogues. The most complete sketch of an astronomical system is contained in the *Timaeus*, but account has to be taken of passages in the *Phaedo*, the *Republic*, the *Laws*, and the *Epinomis*; we have also to make allowance for the admixture of myth and romance which some of the descriptions of astronomical facts contain. At a certain stage Plato seems to have based himself upon the early Pythagorean system (the geocentric system of the Master himself, as distinct from that of the later Pythagoreans who made the earth a planet). We have a glimpse of such a system in the *Republic*. The universe revolves uniformly about its axis, carrying with it all the heavenly bodies; the uniform revolution is that of the daily rotation. At the centre is the earth, immovable, and kept where it is by the equilibrium of symmetry, as it were. The fixed stars are carried round in small circles of the heavenly sphere. The sun, the moon, and the planets are also carried round in the revolution of that sphere, but they have, in addition, circular movements of their own in a sense opposite to that of the daily rotation. Their order in respect of distance from the earth is: moon, sun, Venus, Mercury, Mars, Jupiter, Saturn. The moon describes its own orbit the quickest, the sun the next quickest, while Venus and Mercury travel in company with the sun, each of the three taking about a year to describe its orbit; the next in speed is Mars, the next Jupiter, and the last and slowest Saturn. There is nothing said in the *Republic* about the seven bodies moving in a circle different from and inclined to the equator of the sphere of the fixed stars; that is, the obliquity of the ecliptic does not appear.

In the *Timaeus* the zodiac circle in which the sun, moon, and planets revolve is distinguished as being obliquely inclined to the equator of the sphere of the fixed stars; the latter is called the circle of the Same, the former that of the Other. The other details remain much as described above, except those relating to the earth in the centre. One feature of the system described in the *Timaeus*, namely the fact that the sphere of the fixed stars revolves about its axis and by one revolution produces a night and a day, seems to preclude any rotation of the earth about its axis either in the same or the opposite sense. But the famous passage (40 B) about the earth speaks of it as "rolling" or "winding" (the participle is middle or passive, present tense), and the words "about the pole (axis) stretched through the whole universe," which in the old text came immediately after "rolling," are now, in the text of Burnet, separated from it by the feminine accusative of the article without any substantive agreeing with it. With this article we must assume the accusative of some feminine noun to be understood, and this can hardly be anything but some word for "way," "course," or the like. Given the reading, the expression must admittedly mean "rolling" (or some kind of motion) "on its path about the axis of the universe." Aristotle, in referring to the passage, omits the article, but adds to the verb meaning "to roll" the words "and to move," as if in further elucidation of "rolling," though those words are not in Plato's text; the added words prove that Aristotle took Plato's expression to imply motion of some kind. The latest interpretation of "rolling, etc.," is that of Professor A. E. Taylor, who says we must think, with Burnet, of periodical rectilineal displacements along the axis of the universe in opposite senses and about the centre, and must take the phrase as meaning that the earth moves to and fro, in the sense of sliding up and down the axis of the universe, and so *oscillating* about the centre. Professor Taylor further suggests that Plato is throughout describing a view of Timaeus the Pythagorean which he himself does not share. The objection to the interpretation itself is that Plato does not speak of "rolling *about the centre*" (he does not mention the centre); his words are "about the axis of the universe," and it is hard to see how this can signify a rectilineal displacement *along* the axis. If we are to understand motion "about the axis," I would rather suppose it to be a revolution of the earth in (say) a small orbit round the axis, but such as not to replace or affect in any way the revolution of the sphere of the fixed stars completed in twenty-four hours.

What was Plato's final view on astronomical matters? A change of attitude is certainly indicated in two well-known passages of Plutarch. The first says, with reference to the statement about the earth in the *Timaeus*: "Theophrastus adds that Plato in his later years regretted that he had given the earth the central place in the universe, which was not appropriate to it"; the other passage does not name Theophrastus, but reads: "They say that Plato when old was moved by these considerations [the Pythagorean theory of the central fire] to regard the earth as placed elsewhere than at the centre, and to hold that the central and chiefest place belongs to some worthier body." Now in the passage where Aristotle speaks of the statement about the earth in the *Timaeus* he distinguishes between two theories about the earth's motion. According to him, one of these made the earth move in some way though placed at the centre (this is what he says "is written in the *Timaeus*"); the other took the earth away from the centre and made it move ***round*** the centre. The latter view, he says, was held by the Italian philosophers known as Pythagoreans, who said that "at the centre there is fire, and the earth is one of the stars," etc. But he adds: "Many others might agree that we ought not to assign the central place to the earth, looking for confirmation rather to theory than to observed facts. For they think it appropriate that the most honourable place should belong to the most honourable thing; but fire, they say, is more honourable than earth, and the limit than the intermediate, while the extremity and the centre are both limits. Arguing from these premisses, they think that it is not earth that lies at the centre of the sphere, but rather fire." The verbal resemblances between Aristotle's reference to the most honourable place and the most honourable thing and the two passages of Plutarch suggest that Aristotle may here have Plato in his mind. But can we find in the *Laws* or *Epinomis* any confirmatory evidence that Plato held the view in question? Two passages are relied upon. One is *Laws*, 822 c. The Athenian Stranger there says that it is wrong and even impious to say of the sun, moon, and planets that they never follow the same course but "wander," for in fact "each of them traverses the same path, not many paths, but always one, in a circle, whereas it appears to move in many paths. And again, the swiftest of them is incorrectly thought to be the slowest and vice versa." He further says that this "is not a thing I learnt when I was young or have known a long time." How the scheme is worked is not stated; but one way would be to reject the apparent daily rotation of the fixed stars as an illusion, and to substitute for it either (1) a rotation of the earth itself about its own axis once in twenty-four hours—this was the hypothesis of Heraclides of Pontus—or (2) an equivalent revolution of the earth about the centre of the universe in (say) a small orbit. What light does the *Epinomis* throw on the question? After mentioning seven revolutions, which are those of the sun, moon, and planets, the author says that we must call one revolving body the eighth, namely that which it is most usual to call the universe, and which travels in the opposite sense to the others; he adds, however, that only a person with inadequate knowledge of these things could suppose that this revolution carries the seven others with it. This implies, like the *Laws*, that the revolutions of the sun, moon, and planets are one in each case and not compounded of two or more, though the author seems definitely to retain the revolution of the sphere of the fixed stars. But the latter appears to preclude an axial rotation of the earth, and will only admit of a revolution of the earth in a small orbit round the centre if it is so arranged as not to prejudice or affect the concurrent revolution of the sphere of the fixed stars. If, therefore, we assume that the earth revolves in a small orbit about the centre, we must assume that here, as also in the *Timaeus* (but not in the *Laws*), it is not the earth which makes day and night but the revolution of the sphere of the fixed stars, in other words, while the earth is revolving round the centre, it must *not* incidentally also rotate on its axis (as the moon does in describing its orbit round the earth, since it always turns the same side to the earth). The conclusion of the whole matter seems to be that, while we may quite reasonably suppose Plato, in his later years, to have favoured a hypothesis like that of the Pythagoreans making the earth revolve about the centre like the other planets, it is not possible to find in the *Timaeus* and the later dialogues any clear statement to this effect, or any definite readjustment to this hypothesis of the other details of Plato's system. Perhaps Plato had not arrived at any final opinion, since he is said to have set it as a problem to all earnest students to find "what are the uniform and ordered movements by the assumption of which the motions of the planets can be accounted for."

This no doubt supplied the motive to two important developments in astronomical theory which followed. The first was due to Eudoxus of Cnidos (about 408–355 B.C.), who had early in life attended lectures by Plato, and who brought an extraordinary mathematical insight to bear upon the problem. Eudoxus' hypothesis of concentric spheres, devised for the purpose of explaining the stationary points and retrogradations in the motion of the planets, was the first attempt to furnish a mathematical basis for astronomy, and is a remarkably elegant piece of pure spherical geometry. We are indebted to Aristotle and his commentator Simplicius for what we know of the details. The hypothesis was purely geometrical; there was nothing mechanical about it. For the sun and moon Eudoxus used three spheres in each case; for each of the planets he required four spheres. The spheres were one inside the other, and rotated uniformly about different axes. The revolution of the outermost sphere in each case was that of the daily rotation. The second sphere revolved about an axis perpendicular to the plane of the zodiac or ecliptic, producing motion along the zodiac "in the respective periods in which the planets appear to describe the zodiac circle," i.e. in the case of the superior planets, the sidereal periods of revolution, and in the case of Mercury and Venus (on a geocentric system) one year. The third sphere had its poles at two diametrically opposite points on the zodiac circle, the poles being carried round in the motion of the second sphere. The poles of the third sphere were different for all the planets, except that for Mercury and Venus they were the same. On the surface of the third sphere the poles of the fourth sphere were fixed, and its axis of revolution was inclined to that of the third sphere at an angle constant for each planet but different for the different planets. The planet was fixed at a point on the equator of the fourth sphere. The combined motions of the third and fourth spheres cause the planet to describe, on the surface of the second sphere, the curve called by Eudoxus the "horse-fetter" (*hippopede*), a curve like an elongated figure-of-eight lying along and bisected by the zodiac circle. The motion round this figure-of-eight combined with the motion in the zodiac circle produces the acceleration and retardation of the motion of the planet causing the stations and retrogradations. It was a remarkable feat to prove by pure geometry that this would be the effect of the combined motions; but it was a stroke of genius to invent the combination leading to it.

Eudoxus' system of concentric spheres was further developed by Callippus (about 370–300 B.C.), who added two more spheres for the sun and moon, and one more in the case of each of the planets Venus, Mercury, and Mars. The two additional spheres in the case of the sun were introduced in order to account for the unequal motion of the sun in longitude.

More important astronomically was the second innovation, made by Heraclides of Pontus (about 388–315 B.C.), a pupil of Plato. Heraclides declared in the first place that the apparent daily rotation of the heavenly bodies is due, not to a rotation of the heavenly sphere about an axis through the centre of the earth, but to the rotation of the earth itself about its own axis. Secondly, Heraclides discovered that Venus and Mercury revolve round the sun like satellites. These two capital discoveries constituted important steps towards the Copernican theory; the complete anticipation of Copernicus was reserved for Aristarchus of Samos (about 310–230 B.C.), an elder contemporary of Archimedes.

Before passing to Aristarchus, we have to speak of ARISTOTLE, who was almost contemporary with Heraclides. Aristotle's services to astronomy consist largely of thoughtful criticisms, generally destructive, of opinions held by earlier astronomers; but his own general views of the universe are of interest. According to Aristotle, motion in space is of three kinds; the first is motion in a straight line, the second circular motion, and the third a combination of the two. Of these motions there is only one, namely circular motion, which can be continuous and without beginning or end. Simple bodies have simple motions; thus the four elements tend to move in straight lines; earth tends downwards, fire upwards; between the two are water, the relatively heavy, and air, the relatively light. Hence the order, beginning from the centre, in the sublunary sphere is earth, water, air, fire. Now simple circular motion is more perfect than motion in a straight line. Since therefore there are four elements to which rectilinear motion is natural, there must be another element, different from the four, to which circular motion is natural. This element is superior to the others in proportion to the greater perfection of circular motion and its greater distance from us; it admits of no contraries such as up and down, heavy and light; the absence of contrarieties suggests that it is without beginning or end, imperishable, incapable of increase or change. This element, occupying the uppermost space, is called "aether"; it is a body other and more divine than the four so-called elements; its changelessness is confirmed by long tradition, which contains no record of any alteration in the outer heaven or in any of its proper parts. Of this element are formed the stars, which are spherical, eternal, intelligent, divine. It occupies the whole region from the outside limit of the universe down to the orbit of the moon, though it is not everywhere of uniform purity. It shows the greatest difference where it touches the sublunary sphere. Below the moon is the terrestrial region, the home of the four elements, which is subject to continual change through the strife of those elements and their incessant mutual transformations.

The universe is spherical in shape and is finite, because any body which has a circular motion, as the universe has, must be finite. An infinite body cannot even have a centre about which to rotate. The universe is one only, and complete, containing within it all the matter there is. For all the simple bodies move to their proper places, earth to the centre, aether to the outermost region, and the other elements to the intervening spaces. There can be no simple body outside the universe because that body has its natural place inside. There can be no space or void outside the universe, for space or void is only that in which a body is or can be.

Aristotle was interested, as we have seen, in the purely geometrical hypotheses of concentric spheres put forward by Eudoxus and Callippus, but, in his matter-of-fact way, thought it necessary to transform the system into a mechanical one, with material spherical shells one inside the other and mechanically acting on one another. The object was to substitute one system of spheres for the sun, moon, and planets together, instead of a separate system for each heavenly body. For this purpose he assumed sets of *reacting* spheres between successive sets of the original spheres. Saturn being, for instance, moved by a set of four spheres, he had three reacting spheres to neutralize the last three, in order to restore the outermost sphere to act as the first of the four spheres producing the motion of the next lower planet, Jupiter, and so on. In

Callippus' system there were thirty-three spheres in all; Aristotle added twenty-two reacting spheres, making fifty-five. The change was not an improvement.

ARISTARCHUS of Samos (about 310–230 B.C.) was called "the mathematician," no doubt to distinguish him from the many other persons of the same name; Vitruvius includes him among the few great men who possessed an equally profound knowledge of all branches of science, geometry, astronomy, music, etc. That Aristarchus actually put forward the heliocentric hypothesis is made certain by the evidence of no less a person than Archimedes, who was a younger contemporary of Aristarchus. Archimedes, in a passage of his *Psammites* or *Sand-reckoner*, describes Aristarchus' hypothesis in terms sufficiently precise; the passage will be given in full and need not be repeated here. Plutarch says that Cleanthes the Stoic thought that Aristarchus ought to be indicted on the charge of impiety for "putting in motion the Hearth of the Universe"; and he adds the detail that it was part of Aristarchus' scheme that the earth rotates about its own axis.

Aristarchus found few followers. The only person of note who is mentioned as having argued for Aristarchus' hypothesis is Seleucus of Seleucia, on the Tigris, a Chaldaean who also wrote on the subject of the tides about a century after Aristarchus. Hipparchus, a contemporary of Seleucus, returned to the geocentric system, and no doubt it was his authority which carried the day.

The extant book of Aristarchus, *On the Sizes and Distances of the Sun and Moon*, deserves mention as the first known treatise on the subject which was mathematically worked out on the basis of certain assumptions. These assumptions and some of the conclusions proved will be given later. Other results are quoted in a notice of the work by Pappus of Alexandria which will also be given.

Almost contemporary with Archimedes was ERATOSTHENES of Cyrene, to whom Archimedes dedicated his *Method*. He was a man of great distinction in all branches, though the names Beta and Pentathlos applied to him indicate that he just fell below the first rank in each subject. He was tutor to Ptolemy Euergetes I's son (Philopator), and became librarian at Alexandria after Apollonius Rhodius. His most famous achievement was the measurement of the earth. There are on record two earlier estimates of the size of the earth. According to Aristotle, the mathematicians of his day who had considered the subject made the circumference of the earth to be 400,000 stades. Archimedes says that some persons had tried to prove that it was 300,000 stades. Cleomedes gives a description of Eratosthenes' procedure, as also of the similar but less accurate measurement made by Posidonius the Stoic (about 135–51 B.C.); these accounts will be given in their proper place. Eratosthenes found the circumference of the earth to be 250,000 stades, a figure which he seems later to have corrected to 252,000. Now 252,000 stades, on the most probable assumption as to the length of the stade, are equal to about 24,662 miles, and this gives for the diameter about 7,850 miles, only fifty miles less than the true polar diameter! The figure arrived at by Posidonius for the circumference of the earth was, according to Cleomedes, 240,000 stades.

After Aristarchus, and throughout the centuries down to the time of Copernicus, the theory of the movements of the sun, moon, and planets which found favour was that of epicycles and eccentric circles. Who first formulated the theory of epicycles as such is uncertain; but the idea is inherent in Heraclides' conception of the planets Venus and Mercury revolving round the sun while the sun itself revolves in a circle about the earth; in this case the centre of the epicycle is the material sun. In the general case of an epicycle the planet moves uniformly round the circumference of a circle, the centre of which itself moves uniformly round a larger circle having for its centre the centre of the earth. Motion on an "eccentric" circle is a motion of the same kind; here the planet moves on the eccentric circle and the centre of the eccentric circle simultaneously describes a circle smaller than, and included by, the eccentric circle.

It is certain that APOLLONIUS of Perga (about 265–190 B.C.) discussed the epicycle and eccentric hypotheses generally, and observed that the latter applied to the superior planets only. We may conclude that eccentric circles were invented for the purpose of explaining the movements of Mars, Jupiter, and Saturn about the sun; but it is uncertain who was the inventor, and who was the first to draw the same inference for the superior planets as Heraclides had drawn for the planets Venus and Mercury, and to maintain that all the five planets alike revolved about the sun. This latter hypothesis combined with that of the revolution of the sun about the earth amounted to the system of Tycho Brahe. That system, therefore, as well as that of Copernicus, found a place in Greek thought.

HIPPARCHUS, perhaps the greatest astronomer of antiquity, was born at Nicaea in Bithynia. Ptolemy refers to observations made by him between 161 and 126 B.C., and this indicates roughly the period of his activity. Hipparchus, in investigating the movements of the sun, moon, and planets, adhered to the geocentric system, and proceeded on the alternative hypotheses of epicycles and eccentric circles. While the motions of the sun and moon could with some difficulty be accounted for by the simple epicycle and eccentric hypotheses, Hipparchus found that for the planets it was necessary to combine the two, that is, to superadd epicycles to motion in eccentric circles. Hipparchus is the first person who is known to have made systematic use of trigonometry in his work. He compiled a Table of Chords in a circle, which is equivalent to a table of trigonometric sines. He made great improvements in the instruments used for astronomical observations. He compiled a catalogue of fixed stars to the number of 850 or more, and he appears to have been the first to state their positions in terms of co-ordinates in relation to the ecliptic (latitude and longitude).

One of Hipparchus' greatest achievements was the discovery of the Precession of the Equinoxes. He calculated its amount to have been two degrees in the period between his own observation of the bright star Spica and that recorded by Timocharis 154 or 166 years earlier; this gives, according as we take one or other of these figures, a rate of about 46·8 seconds or 43·4 seconds, as compared with the true rate of 50·3757 seconds.

In a tract, *On the Length of the Year*, Hipparchus

calculated that the tropic year is nearly one-three-hundredth part of a day and a night less than the previously accepted figure of $365\frac{1}{4}$ days; this works out to 365 days, 5 hours, 55 minutes, 12 seconds, which exceeds the true mean tropic year by about $6\frac{1}{2}$ minutes. From Hipparchus' estimate of the "Great Year" (304 years including 112 intercalary months) we obtain as the length of the mean lunar month 29·530585 days, or 29 days, 12 hours, 44 minutes, and $2\frac{1}{2}$ seconds, which is less than a second out in comparison with the present accepted figure of 29·530596 days! It is right to add that Hipparchus must have had access to Babylonian calculations and in particular to those of Naburiannu (flourished about 500 B.C.) and Kidinnu (about 383 B.C.), which work out to 29·530614 and 29·530594 days respectively.

Lastly, Hipparchus improved on Aristarchus' estimates of the sizes and distances of the sun and moon; he noted the changes in their apparent diameters and made the mean distances of the sun and moon to be $1245D$ and $33\frac{2}{3}D$, and their diameters $12\frac{1}{3}D$ and $\frac{1}{3}D$ respectively, where D is the mean diameter of the earth.

The astronomy of Hipparchus takes its definitive form in the *Syntaxis* (commonly called the *Almagest*) of Ptolemy, written about A.D. 150, which held the field till the time of Copernicus. It is questionable whether Ptolemy himself added anything of great value except a definite theory of the motion of the five planets, for which Hipparchus had only collected material in the shape of observations made by his predecessors and himself.

Such being, in brief outline, the story of Greek astronomy, we naturally ask what it was that the Greeks brought to the study of the subject which gave to their astronomy its special character in comparison with that of Egypt and Babylon. The answer is, first, an unrivalled speculative genius, and secondly—a particular manifestation of the same thing—their mathematics. Recent studies of Babylonian sources have shown that we must revise former estimates of the extent to which the Greeks were indebted for the details of their astronomy to the Babylonians; the debt proves to have been much greater than had been imagined, and further researches may prove it to have been greater still. But, so far as we can judge, the Greeks were original and independent (1) in their cosmological speculations and (2) in their *theories* of astronomy. Thales is, in his views of the universe and its origin, on the level of contemporary or earlier Babylonian thought. But, when we come to Anaximander, Anaxagoras, Leucippus, and Democritus, we are in a new realm. Their theories, audacious as they were for their age, contain elements which will be fruitful for all time; we have only, in illustration of this, to think of the points in which they show a certain affinity with the most modern cosmogonies on the one hand, and with the recent developments in the theory of the atom on the other. If the Babylonians also put forward physical theories of the same sort, no trace of them seems so far to have been found.

In the domain of astronomy the Egyptians defined the year, gave names to the months, divided night and day into twelve hours each, invented water-clocks and a species of sun-clock. The Babylonians went further, distinguished stars according to groups, including the signs of the zodiac, invented their own types of sundials, devised cycles for the prediction of eclipses and for intercalations of months, and introduced the sexagesimal system of calculation; they could draw plane representations of the circles in the heaven; they observed and recorded the apparent motions of the planets. But did they formulate any theory to account for the movements of the sun, moon, and planets? Apparently not, so far as we know. It was here that the Greek mathematical genius came in. The first purely mathematical theory was Eudoxus' system of concentric spheres, a marvel of geometrical acumen; the geometry of the sphere and the circles in it had already been fully worked out, possibly by Eudoxus himself. About the same time Heraclides of Pontus saw that the apparent daily revolution of the heavenly sphere about the earth as centre could be replaced by a rotation of the earth about its own axis in twenty-four hours; he declared also that Venus and Mercury revolve about the sun like satellites. Heraclides thus made a great advance towards the system of epicycles on the one hand, and towards the Copernican hypothesis on the other. The Copernican hypothesis was actually anticipated by Aristarchus of Samos; and a system precisely the same as that of Tycho Brahe is also represented in Greek astronomy, whether it was Apollonius or Perga or another who formulated it. Finally the geometrical hypothesis of epicycles and eccentric circles, which took account of the variations in the distances of the sun, moon, and planets at different times, was elaborated and tested. It was Kepler that the Greeks failed to anticipate.

EPIGRAM

I KNOW that I am mortal and the creature of a day; but when I search out the massed wheeling circles of the stars, my feet no longer touch the earth, but, side by side with Zeus himself, I take my fill of ambrosia, the food of the gods.

PTOLEMY

THALES

PLATO, *Theaetetus*, 174 A.

A CASE in point is that of Thales, who, when he was star-gazing and looking upward, fell into a well, and was rallied (so it is said) by a clever and pretty maidservant from Thrace because he was eager to know what went on in the heaven, but did not notice what was in front of him, nay, at his very feet.

HERODOTUS, I, 74.

When, in the sixth year, they [the Lydians and the Medes] encountered one another, it so fell out that, after they had joined battle, the day suddenly turned into night. Now that this transformation of day (into night) would occur was foretold to the Ionians by Thales of Miletus, who fixed as the limit of time this very year in which the change actually took place.

CLEMENT OF ALEXANDRIA, *Stromat.* I, 65.

Eudemus observes in his *History of Astronomy* that Thales predicted the eclipse of the sun which took place at the time when the Medes and the Lydians engaged in battle, the King of the Medes being Cyaxares, the father of Astyages, and the King of the Lydians being Alyattes, the son of Croesus; and the time was about the fiftieth Olympiad [580–577 B.C.]. (But cf. Pliny, *N. H.*, c. 12, § 53: "Among the Greeks Thales was the first of all men to investigate (the cause of eclipses), in the fourth year of the forty-eighth Olympiad [585/4 B.C.], he having predicted an eclipse of the sun which took place in the reign of Alyattes in the year 170 A.U.C.")

THEON OF SMYRNA, p. 198.

Eudemus relates in his *Astronomies* . . . that Thales was the first to discover the eclipse of the sun, and the fact that the sun's period with respect to the solstices is not always the same.

DIOGENES LAËRTIUS, I, 24.

Thales was the first to discover the length of the interval from solstice to solstice.

ARISTOTLE, *Metaph.* A. 3, 983 b 20–2.

Thales, the originator of this kind of philosophical inquiry [i.e. the search for one material cause of all things], says that water is the first principle (this is why he also declared that the earth rests on water).

DIOGENES LAËRTIUS, I, 27.

Thales laid it down that the first principle of all things is water, and that the universe is animate and full of gods. They say too that he discovered the seasons of the year, and divided the year into 365 days.

HERODOTUS, II, 4.

It is said that the Egyptians were the first of all men to discover the year, to which they gave twelve parts (months) making up the (four) seasons. And herein the Egyptians reckon, as it seems to me, more sensibly than the Greeks, in so far as the Greeks put in an intercalary month every third year to keep the seasons right, whereas the Egyptians reckon their twelve months at thirty days each and add in every year five days outside the number (of twelve times thirty).

DIOGENES LAËRTIUS, I, 23.

Callimachus knows him (Thales) to be the discoverer of the "Little Bear," for he says in his *Iambi* that "he was said to have marked (*lit.* 'measured') the stars of the Wain, by which the Phoenicians sail."

ANAXIMANDER

SIMPLICIUS, in *Phys. Aristotelis*, p. 24, 13, Diels.

ANAXIMANDER of Miletus, who was a fellow-citizen and friend of Thales, said that the first principle (i.e. material cause) and element of existing things is the Infinite, and he was the first to introduce this name for the first principle. He maintains that it is neither water nor any other of the so-called elements, but another sort of substance which is infinite, and from which all the heavens and the worlds in them are produced; and into that from which existent things arise they pass away once more, "as is ordained; for they must pay the penalty and make reparation to one another for the injustice they have committed, according to the sequence of time," as he says in these somewhat poetical terms.

PSEUDO-PLUTARCH, *Stromat.* 2.

Anaximander said that the Infinite contains the whole cause of the generation and destruction of the All; it is from the Infinite that the heavens are separated off, and generally all the worlds, which are infinite in number. He declared that destruction and, long before that, generation have been going on from infinitely distant ages, all the worlds recurring in cycles.

HIPPOLYTUS, *Refutation of all Heresies*, I, 6.

He says that this [the Infinite] is eternal and ageless and embraces all the worlds. And he implies the existence of time in that the three stages of coming into being, existence, and passing-away are distinguished. . . . And besides the Infinite he says there is eternal motion, in the course of which it happens that the heavens come into being.

HERMIAS, *Irrisio*, 10.

Anaximander says eternal motion is a principle older than the moist, and it is by this eternal motion that some things are generated and others destroyed.

AËTIUS, *De placitis*, I, 3, 3.

He says that the first principle (or material cause) is infinite, in order that the process of coming into being which is set up may not suffer any check.

SIMPLICIUS, on *De caelo*, p. 615, 13, Heiberg.

Anaximander was the first to assume the Infinite as first principle, in order that he might have it available for his new births without stint.

SIMPLICIUS, in *Phys.* p. 1121, 5.

Those who assumed that the worlds are infinite in number, as did Anaximander, Leucippus, Democritus, and, in later days, Epicurus, assumed that they also came into being and passed away, *ad infinitum*, there being always some worlds coming into being and others passing away; and they maintained that motion is eternal: for without motion there is no coming into being or passing away.

PSEUDO-PLUTARCH, *Stromat.* 2.

Anaximander says that that which is capable of begetting the hot and the cold out of the eternal was separated off during the coming into being of our world, and from this there was produced a sort of sphere of flame which grew round the air about the earth as the bark round a tree; then this sphere was torn off and became enclosed in certain circles or rings, and thus were formed the sun, the moon, and the stars.

HIPPOLYTUS, *Refut.* I, 6, 4, 5.

The stars are produced as a circle of fire, separated off from the fire in the universe and enclosed by air. They have as vents certain pipe-shaped passages at which the stars are seen; hence, when the vents are stopped up, eclipses take place. The moon appears sometimes as waxing, sometimes as waning, to an extent corresponding to the closing or opening of the passages.

AËTIUS, II, 13, 7.

The stars are compressed portions of air, in the shape

of wheels filled with fire, and they emit flames at some point from small openings.

HIPPOLYTUS, loc. cit.

The earth is poised aloft, supported by nothing, and remains where it is because of its equidistance from all other things. Its form is rounded, circular, like a stone pillar; of its plane surfaces one is that on which we stand, the other is opposite.

PSEUDO-PLUTARCH, *Stromat.* 2.

The earth, he says, is cylinder-shaped and its depth is such as to have a ratio of one-third to its breadth.

HIPPOLYTUS, *Refut.* I, 6, 5.

The circle of the sun is twenty-seven times as large <as the earth, and that> of the moon <is nineteen times as large as the earth>.

AËTIUS, II, 20, 21, 24, 25, 29, 15.

(20, 1) The sun is a circle twenty-eight times the size of the earth; it is like a chariot-wheel, the rim of which is hollow and full of fire, and lets the fire shine out at a certain point in it through an opening like the nozzle of a pair of bellows: such is the sun.

(21, 1) The sun is equal to the earth, and the circle from which the sun gets its vent and by which it is borne round is twenty-seven times the size of the earth.

(24, 2) The eclipses of the sun occur through the orifice by which the fire finds vent being shut up.

(25, 1) The moon is a circle nineteen times as large

as the earth; it is like a chariot-wheel, the rim of which is hollow and full of fire, like the circle of the sun, and it is placed obliquely, as that of the sun also is; it has one vent like the nozzle of a pair of bellows; its eclipses depend on the turnings of the wheel.

(29, 1) The moon is eclipsed when the orifice in the rim of the wheel is stopped up.

(15, 6) The sun is placed highest of all, after it the moon, and under them the fixed stars and the planets.

EUSEBIUS, *Praeparatio evangelica*, X, 14, 11.

Anaximander was the first to construct *gnomons* (sundials) for the purpose of distinguishing the turnings-back of the sun (solstices), times, seasons, and the equinoxes. (But cf. Herodotus, II, 109: "The Greeks learnt from the Babylonians the use of the *polos* and the *gnomon*, and also the twelve parts of the day.")

DIOGENES LAËRTIUS, II, 2.

He was the first to make a drawing of the contour of the (inhabited) earth and the sea.

AGATHEMERUS, I, 1.

Anaximander of Miletus, a pupil of Thales, was the first who ventured to make a drawing of the inhabited earth on a tablet, and after him Hecataeus of Miletus, a much-travelled man, corrected it so that it became an object of general admiration.

HIPPOLYTUS, *Refut.* I, 6, 6.

(Anaximander held) that living creatures arose <from the moist element> evaporated by the sun. Man was in the beginning like another living creature, namely a fish.

Aëtius, V, 19, 4.

Anaximander said that the first living creatures came into existence in the moist element, and had prickly coverings, but, as they advanced in age, they moved to the drier part; and, when the covering peeled off, they survived in their changed state for a short time.

Plutarch, *Symp.* VIII, 8, 4.

The descendants of Hellen of old sacrifice to the ancestral Poseidon, thinking, like the Syrians, that man was born of the moist substance; hence they venerate the fish as being of like race and like nurture with man, a doctrine which is more reasonable than that of Anaximander; for the latter declares, not that fishes and men lived in the same conditions, but that men were at first born inside fishes, and were reared like sharks, after which, when they became capable of fending for themselves, they left the water and took to land.

Pseudo-Plutarch, *Stromat.* fr. 2.

Anaximander says that at the beginning man was born from other species of animals; this he inferred from the fact that, while other animals quickly manage to find food for themselves, man alone needs long nursing. If he had been what he is now, he could not possibly have survived.

ANAXIMENES

Simplicius, in *Phys.* p. 24, 26.

Anaximenes of Miletus, son of Eurystratus, who had been an associate of Anaximander, said, like him, that the underlying substance was one and infinite. He did not, however, say it was indeterminate, like Anaximander, but determinate; for he said it was Air. It differs in different substances in virtue of its rarefaction and condensation.

Hippolytus, *Refut.* I, 7.

From it (i.e. Air), he said, the things that are, and have been, and shall be, the gods and things divine, took their rise, while other things came from its offspring. And the form of the air is as follows. Where it is most even, it is invisible to our sight; but cold and heat, moisture and motion, make it visible. It is always in motion; for, if it were not, it would not change so much as it does. When it is dilated so as to be rarer, it becomes fire; while winds, on the other hand, are condensed Air. Cloud is formed from Air by felting; and this, still further condensed, becomes water. Water, condensed still more, turns to earth; and when condensed as much as it can be, to stones.

Aëtius, I, 3, 4.

"Just as," he said, "our soul, being air, holds us together, so do breath and air encompass the whole world."

PSEUDO-PLUTARCH, fr. 3.

Anaximenes says that, as the air was felted, the earth first came into being. It is very broad and is accordingly supported by the air.

HIPPOLYTUS, loc. cit.

In the same way, the sun and the moon and the other heavenly bodies, which are of a fiery nature, are supported by the air because of their breadth. The heavenly bodies were produced from the earth by moisture rising from it. When this is rarefied, fire comes into being, and the stars are composed of the fire thus raised aloft. There were also bodies of earthy substance in the region of the stars, revolving along with them. And he says that the heavenly bodies do not move under the earth, as others suppose, but round it, as a cap turns round our head. The sun is hidden from sight, not because it goes under the earth, but because it is concealed by the higher parts of the earth, and because its distance from us becomes greater. The stars give no heat because of the greatness of their distance.

AËTIUS, II, 14, 16, 20, 22, 23, 25; III, 10.

(II, 14, 3) The stars [are fixed like nails in the crystalline vault of the heavens, but some say they] are fiery leaves, like paintings.

(16, 6) They do not go under the earth, but turn round it.

(20, 2) The sun is fiery.

(22, 1) It is broad like a leaf.

(23, 1) The heavenly bodies turn back in their courses owing to the resistance of compressed air.

(25, 2) The moon is of fire.

(III, 10, 3) The earth was like a table in shape.

JOHN BURNET.

PYTHAGORAS

DIOGENES LAËRTIUS, VIII, 48.

FURTHER we are told that Pythagoras was the first to call the heaven the universe, and the earth round (i.e. spherical), though according to Theophrastus it was Parmenides, and according to Zeno it was Hesiod.

Ib., IX, 23.

And he (Parmenides) is thought to have been the first to see that the Evening Star and the Morning Star are one and the same, as Favorinus says in the fifth book of his *Memorabilia*; but others say it was Pythagoras. (Cf. VIII, 14: "It was he (Pythagoras) who first declared that the Evening and Morning Stars are the same, as Parmenides maintains.")

THEON OF SMYRNA, p. 150, 12–18.

The impression of variation in the movement of the planets is produced by the fact that they appear to us to be carried through the signs of the zodiac in certain circles of their own, being fastened in spheres of their own and moved by their motion, as Pythagoras was the first to observe, a certain varied and irregular motion being thus grafted, as a qualification, upon their simply and uniformly ordered motion in one and the same sense (i.e. that of the daily rotation from east to west).

Aëtius, I, 21, 1.

Pythagoras held that time is the sphere of the enveloping (heaven).

Aristotle, *Phys.* Δ 10, 218 a 33.

Some (of the Pythagoreans) say that time is the motion of the whole (the universe), others that it is the sphere itself.

Ib., Δ 6, 213 b 22–4.

The Pythagoreans also held that void exists, and that the void enters the heaven itself from the infinite breath (outside it), as if the heaven inhaled even the void.

ALCMAEON

Aëtius, II, 16, 2–3.

Alcmaeon and the mathematicians hold that the planets have a motion from west to east, in a direction opposite to that of the fixed stars.

XENOPHANES

ARISTOTLE, *Metaph.* A 5, 986 b 21–4.

XENOPHANES was the first of these philosophers to maintain the doctrine of the One, though he made no clear statement on the subject . . . but, referring to the whole Heaven, he states that the One is God.

Fragments, 11, 13–29 (tr. JOHN BURNET).

Homer and Hesiod have ascribed to the gods all things that are a shame and a disgrace among mortals, stealings and adulteries and deceivings of one another.

But mortals deem that the gods are begotten as they are, and have clothes like theirs, and voice and form.

Yes, and if oxen and horses or lions had hands, and could paint with their hands, and produce works of art as men do, horses would paint the forms of the gods like horses, and oxen like oxen, and make their bodies in the image of their several kinds.

The Ethiopians make their gods black and snub-nosed; the Thracians say theirs have blue eyes and red hair.

The gods have not revealed all things to men from the beginning, but by seeking they find in time what is better.

One god, the greatest among gods and men, neither in form like unto mortals nor in thought.

He sees all over, thinks all over, and hears all over.

But without toil he swayeth all things by the thought of his mind.

And he abideth ever in the selfsame place, moving not

at all; nor doth it befit him to go about now hither, now thither.

All things come from the earth, and in earth all things end.

The limit of the earth above is seen at our feet in contact with the air; below it reaches down without a limit.

All things are earth and water that come into being and grow.

THEODORETUS, IV, 5.

Xenophanes, conceiving the whole to be one, declared it to be spherical, finite, unoriginated, but eternal and altogether without motion; on the other hand he forgot these statements when he said that everything was produced from the earth.

AËTIUS, II, 13, 14; 20, 3; 24, 4, 9.

The stars are made of clouds set on fire; extinguished every day, they are rekindled at night like coals; their risings and settings are lightings and extinguishings respectively.

The sun is made of clouds set on fire. Theophrastus in his *Physics* has used the words "of fiery particles collected together from moist exhalation," and describes the sun as one of the things which collect them.

Xenophanes says that (the setting of the sun happens) by way of extinguishment, and that another sun appears at its rising again.

There are many suns and moons according to the regions, divisions, and zones of the earth, and at certain times the disk falls upon some division of the earth not inhabited by us, and thus when, as it were, stepping where there is void, exhibits eclipse. The same philosopher says that the sun goes forward to infinity, though it appears to revolve in a circle owing to its distance.

HIPPOLYTUS, *Refut.* I, 14, 15.

Xenophanes says that a mixture of the earth and the sea seems to take place and in course of time is dissolved by the moisture. For this he says he has the following proofs. Shells are found far inland and on mountains, and he tells us that in the quarries at Syracuse imprints (fossils) of fish and seaweed have been found, and at Paros an imprint of a bayleaf (reading δάφνης) in the depth of the stone, and at Malta impressions of all marine creatures. These things, he says, were produced at a time long ago when all things were turned into mud, and the impression was dried in the mud. And men must all perish when the earth has been carried down into the sea and become mud: and from that point again generation begins, and this change takes place in all the worlds.

HERACLITUS

DIOGENES LAËRTIUS, IX, 8–11.

HERACLITUS' opinions on particular points are these:

He held that Fire was the element, and that all things were an exchange for fire, produced by condensation and rarefaction. But he explains nothing clearly. All things were produced in opposition, and all things were in flux like a river.

The all is finite and the world is one. It arises from fire, and is consumed again by fire alternately through all eternity in certain cycles. This happens according to fate. Of the opposites, that which leads to the becoming of the world is called War and Strife; that which leads to the final conflagration is Concord and Peace.

He called change the upward and the downward path, and held that the world comes into being in virtue of this. When fire is condensed, it becomes moist, and when compressed it turns to water, water being congealed turns to earth, and this he calls the downward path. And, again, the earth is in turn liquefied, and from it water arises, and from that everything else; for he refers almost everything to the evaporation from the sea. This is the path upwards.

He held, too, that exhalations arose both from the sea and the land; some bright and pure, others dark. Fire was nourished by the bright ones, and moisture by the others.

He does not make it clear what is the nature of that which surrounds the world. He held, however, that there were bowls in it with the concave sides turned towards us, in which the bright exhalations were collected and produced flames. These were the heavenly bodies.

The flame of the sun was the brightest and warmest; for the other heavenly bodies were more distant from the earth, and for that reason gave less light and heat. The moon, on the other hand, was nearer the earth; but it moved through an impure region. The sun moved in a bright and unmixed region, and at the same time was at just the right distance from us. That is why it gives more heat and light. The eclipses of the sun and moon were due to the turning of the bowls upwards, while the monthly phases of the moon were produced by a gradual turning of the bowl.

Day and night, months and seasons and years, rains and winds, and things like these, were due to the different exhalations. The bright exhalation, when ignited in the circle of the sun, produced day, and the preponderance of the opposite exhalations produced night. The increase of warmth proceeding from the bright exhalation produced summer, and the preponderance of moisture from the dark exhalation produced winter. He assigns the causes of other things in conformity with this.

As to the earth, he makes no clear statement about its nature, any more than he does about that of the bowls.

JOHN BURNET.

PARMENIDES

Fragments.

(8) One path only is left for us to speak of, namely that *It is.* In this path are very many tokens that what is is uncreated and indestructible; for it is complete, immovable, and without end. Nor was it ever, nor will it be; for now *it is*, all at once, a continuous one. . . .

Moreover it is immovable in the bonds of mighty chains, without beginning and without end; since coming into being and passing away have been driven afar, and true belief has cast them away. It is the same, and it rests in the selfsame place, abiding in itself. And thus it remaineth constant in its place; for hard necessity keeps it in the bonds of the limit that holds it fast on every side.

Wherefore it is not permitted to what is to be infinite; for it is in need of nothing; while, if it were infinite, it would stand in need of everything. . . .

Since, then, it has a furthest limit, it is complete on every side, like the mass of a rounded sphere, equally poised from the centre in every direction; for it cannot be greater or smaller in one place than another. . . .

Here shall I close my trustworthy speech and thought about the truth. Henceforward learn the beliefs of mortals, giving ear to the deceptive ordering of my words. . . .

(9) Now that all things have been named light and night, and the names which belong to the power of each have been assigned to these things and to those, everything is full at once of light and dark night, both equal, since neither has aught to do with the other.

(10, 11) And thou shalt know the substance of the sky, and all the signs in the sky, and the resplendent works of the glowing sun's pure torch, and whence they arose. And thou shalt learn likewise of the wandering deeds of the round-faced moon, and of her substance. Thou shalt know, too, the heavens that surround us, whence they arose, and how Necessity took them and bound them to keep the limits of the stars . . . how the earth, and the sun, and the moon, and the sky that is common to all, and the Milky Way, and the outermost Olympus, and the burning might of the stars arose.

(12) The narrower bands were filled with unmixed fire, those next them with night, and in the midst of these rushes their portion of fire. In the midst of these is the divinity that directs the course of all things.

JOHN BURNET.

Aëtius, II, 7, 1.

Parmenides held that there are certain wreaths (or bands) twined round, across one another; one sort is made of the rarefied (element), the other of the condensed; and between these are others consisting of light and darkness in combination. That which encloses them all is solid, like a wall, below which is a wreath of fire; that which is in the very middle of all the wreaths is solid, and about it again is a wreath of fire. And of the mixed wreaths, the midmost is to all of them the beginning and cause of motion and becoming, and this he called the "Deity which directs their course," the "Holder of Lots," "Justice" and "Necessity." Moreover, the "air" is thrown off the earth in the form of vapour, owing to the violent pressure of its condensation; the sun and the Milky Way are an exspiration of the fire. The moon is a mixture of both elements, "air" and fire. And, while the encircling aether is uppermost of all, below it is ranged that fiery (thing) which we call heaven, under which, again, are the regions round the earth.

Ib., III, 1, 4.

It is the mixture of the dense and the rarefied which produces the colour of the Milky Way.

Ib., II, 20, 8 a.

The sun and the moon were separated off from the Milky Way, the sun arising from the more rarefied mixture, which is hot, and the moon from the denser, which is cold.

Ib., II, 15, 4.

Parmenides places the Morning Star, which he thinks the same as the Evening Star, first in the aether; then, after it, the sun, and under it again, the stars in the fiery (thing) which he calls heaven.

Ib., II, 26, 2.

The moon Parmenides declared to be equal to the sun; for indeed its illumination comes from it.

Fragments 14, 15.

"A night-shining foreign light wandering round the earth."

"(The moon) always fixing its gaze on the beams of the sun."

Aëtius, III, 15, 7.

Parmenides and Democritus maintain that it is because it is equidistant (from the extremities) in all directions that the earth remains in equilibrium, having no reason why it should incline this way rather than that.

EMPEDOCLES

Aëtius, II, 6, 3.

Empedocles said that aether (air) was first separated out, and secondly, fire. After this came earth, from which, as it was being excessively compressed by the force of the revolution, water gushed forth. From water "air" (mist) was produced by evaporation, and the heaven was formed from the air, and the sun from the fire, while the things about the earth were formed by felting from the other elements.

PSEUDO-PLUTARCH, *Stromat.* fr. 10.

Empedocles held that the elements were four: fire, water, aether (air), and earth. And Love and Strife are the causes of these. From the first mixture of the elements Air was separated out and was spread round in a circle. After the Air, Fire, running out and finding no other place, ran out upwards under the solid enclosing the air. There are, revolving in a circle round the earth, two hemispheres, one of which is wholly of fire, and the other is a mixture of air and a little fire: the latter he supposes to be night. The beginning of the revolution came about through the collection having taken place (in one hemisphere), the fire having preponderated (and so upset the equilibrium).

Fragment 48.

It is the earth which makes night to its lights (by obstructing them) when it (the sun) goes underneath.

AËTIUS, II, 11, 2.

The heaven is solid and made of air (mist) congealed by fire, like crystal, and encloses the fiery and air-like (contents) of the two hemispheres respectively.

Ib., II, 31, 4.

The height from the earth to the heaven, that is, its (vertical) distance from us, is exceeded by its lateral dimension, the heaven having opened out more in this sense, owing to the universe being like an egg in shape.

PSEUDO-PLUTARCH, *Stromat.* fr. 10.

The sun is, in its nature, not fire, but a reflection of fire similar to that which takes place from (the surface of) water.

AËTIUS, II, 20, 13.

Empedocles says there are two suns: one is the original sun which is the fire in one hemisphere of the world, filling the whole hemisphere and always placed directly opposite to the reflection of itself; the other is the apparent sun, which is a reflection in the other hemisphere filled with air and an admixture of fire, and, in this reflection, what happens is that the light is bent back from the earth, which is circular, and is concentrated into the crystalline sun, where it is carried round by the motion of the fiery (hemisphere). Or, to state the fact shortly, the sun is a reflection of the fire about the earth.

Ib., II, 21, 2.

The sun which consists of the reflection is equal in size to the earth.

PLUTARCH, *De Pyth. Or.*, 12, p. 400 B.

You laugh at Empedocles for saying that the sun is produced about the earth by a reflection of the light in the heaven, and "once more flashes back to Olympus with fearless countenance."

Pseudo-Plutarch, *Stromat.* fr. 10.

The moon, he says, was composed as a separate body out of the air cut off by the fire. For this air froze just like hail. And it has its light from the sun.

Diogenes Laërtius, VIII, 77.

The sun, he says, is a vast collection of fire, and larger than the moon; the moon is disk-shaped, and the heaven itself is crystalline.

Aëtius, II, 31, 1.

Empedocles said that the moon is twice as far distant from the sun as it is from the earth.

Fragment 42.

"The moon shuts off the beams of the sun as it passes across it, and darkens so much of the earth as the breadth of the blue-eyed moon amounts to."

Aëtius, II, 13, 2.

The stars are of fire (arising) out of the fiery element which the air contained in itself, but squeezed out upwards in the original separation.

Ib., II, 13, 11.

The fixed stars are attached to the crystal sphere; the planets are free.

Ib., II, 1, 4.

The sun's course is the circuit of the limit of the world.

Ib., II, 23, 3.

Empedocles said that the sun is prevented from moving always in a straight line by the sphere enveloping it and by the tropic circles.

Light travels

Aristotle, *De sensu*, c. 6, 446 a 26–b 2.

Empedocles, for instance, says that the light from the sun reaches the intervening space before it reaches the eye or the earth. And this might well seem to be the fact. For, when a thing is moved, it is moved from one place to another, and hence a certain time must elapse during which it is being moved from the one place to the other. But every period is divisible. Therefore, there was a time when the ray was not seen, but was being transmitted through the medium.

Aristotle, *De anima*, II, 7, 418 b 21–23.

Empedocles represented light as moving in space, and arriving at a given point of time between the earth and that which surrounds it, without our perceiving its motion.

ANAXAGORAS

Fragments.

(1) All things were together, infinite both in number and in smallness; for the small, too, was infinite. And, when all things were together, none of them could be distinguished for their smallness. For air and aether prevailed over all things, being both of them infinite;

for amongst all things these are the greatest both in quantity and size.

(11) In everything there is a portion of everything except Nous, and there are some things in which there is Nous also.

(12) All other things partake in a portion of everything, while Nous is infinite and self-ruled, and is mixed with nothing, but is alone, itself by itself. . . . And Nous has power over all things, both greater and smaller, that have life. And Nous had power over the whole revolution, so that it began to revolve in the beginning. And it began to revolve first from a small beginning; but the revolution now extends over a larger space, and will extend over a larger still. And all the things that are mingled together and separated off and distinguished are known by Nous. And Nous set in order all things that were to be, and all things that were and are not now, and that are, and this revolution in which now revolve the stars and the sun and the moon, and the air and the aether that are separated off. And this revolution caused the separating off, and the rare is separated off from the dense, the warm from the cold, the light from the dark, and the dry from the moist. . . .

(15) The dense and the moist and the cold and the dark came together where the earth is now, while the rare and the warm and the dry (and the bright) went out towards the further part of the aether.

(16) From these as they are separated off earth is solidified; for from mists water is separated off; and from water, earth. From the earth stones are solidified by the cold, and these rush outwards more than water.

John Burnet.

Aëtius, II, 13, 3.

Anaxagoras maintained that the surrounding aether is fiery in substance, and that in consequence of the strength of the whirling motion it tore away stones from the earth, and, setting them on fire, kindled them into stars.

Hippolytus, *Refut.* I, 8.

The earth is flat in shape, and remains suspended because of its size, because there is no void, and because the air, being very strong, supports the earth which rides upon it.

The sun, the moon, and all the stars are stones on fire, which are carried round by the revolution of the aether. And there are, below the stars, the sun and the moon, certain other bodies carried round with them, but invisible to us.

We do not feel the heat of the stars because they are at a great distance from the earth; besides which they are not as hot as the sun because they occupy a colder region. The moon is below the sun and nearer to us.

The sun exceeds the Peloponnesus in size [Aëtius has "is many times larger than the Peloponnese"]. The light which the moon has is not its own but comes from the sun. The revolution of the stars takes them under the earth.

The moon is eclipsed through the interposition of the earth, and sometimes also of the bodies below the moon. The sun is eclipsed at the new moon, when the moon is interposed. The sun and the moon execute their turnings-back owing to their being thrust back by the air. The moon's turnings are frequent because it cannot get the better of the cold.

Anaxagoras was the first to set out distinctly the facts about eclipses and illuminations. He declared that the moon is of earthy nature, and has in it plains and ravines. The Milky Way is the reflection of the light of the stars which are not illuminated by the sun. Shooting-stars are sparks, as it were, which are made to leap out, owing to the motion of the heavenly sphere.

PLATO, *Cratylus*, 409 A.

"The fact which he (Anaxagoras) recently asserted, namely that the moon has its light from the sun."

PLUTARCH, *De facie in orbe lunae*, 16, p. 929 B.

Now when our comrade, in his discourse, had expounded that proposition of Anaxagoras that "the sun places the brightness in the moon," he was greatly applauded.

AËTIUS, II, 29, 6.

Anaxagoras, in agreement with the mathematicians, held that the moon's obscurations, month by month, were due to its following the course of the sun by which it is illuminated, and that the eclipses of the moon were caused by its falling within the shadow of the earth, which then comes between the sun and the moon, while the eclipses of the sun were due to the interposition of the moon.

Ib., II, 29, 7.

Anaxagoras, as Theophrastus says, held that the moon was also sometimes eclipsed by the interposition of the bodies below the moon.

Ib., II, 25, 9.

The moon is an incandescent solid, having in it plains, mountains, and ravines.

Ib., II, 30, 2.

It is an irregular compound because it has an admixture of cold and of earth. It has a surface in some places lofty, in others low, in others hollow. And the dark is mixed together with the fiery, the joint effect being an impression of the shadowy; hence it is that the moon is said to shine with a false light.

The Milky Way

Aristotle, *Meteor.* A 8, 345 a 25–31.

Anaxagoras and Democritus held that the Milky Way is the light of certain stars. For, when the sun is passing below the earth, some of the stars are not within its vision. Such stars, then, as are embraced in its view are not seen to give light, for they are overpowered by the rays of the sun; such of the stars however as are hidden by the earth, so that they are not seen by the sun, form, by their own proper light, the Milky Way.

Plutarch, *Nicias*, 23.

For Anaxagoras, who was the first to put in writing, most clearly and most courageously of all men, the explanation of the moon's illumination and darkness, did not belong to ancient times, and even his account was not common property, but was still a secret, current only among a few, and received by them with caution, or simply on trust. For in those days they refused to tolerate the natural philosophers and star-gazers, as they were then called, who presumed to fritter away the deity into unreasoning causes, blind forces, and necessary properties. Thus Protagoras was exiled, and Anaxagoras was imprisoned, and with difficulty saved by Pericles.

Other worlds than ours

Fragments 4.

Men have been formed and the other animals which have life; the men, too, have inhabited cities and cultivated fields as with us; they have a sun and moon and the rest as with us, and their earth produces for them many things of various kinds, the best of which they gather together into their dwellings and live upon. Thus much have I said about separating off, to show that it will not be only with us that things are separated off, but elsewhere as well.

Aëtius, II, 8, 1.

Diogenes (of Apollonia) and Anaxagoras held that, after the world was formed, and the animals were produced from the earth, the world received, as it were, an automatic tilt towards its southern part, perhaps by design, in order that some parts of the world might become uninhabitable and others habitable, according as they are subject to extreme cold, torrid heat, or moderate temperature.

THE PYTHAGOREANS

Aristotle, *De caelo*, B 13, 293 a 15–b 30.

It remains to speak of the earth, its position, the question whether it belongs to the class of things which are at rest, or to that of things in motion, and its shape. Regarding its position opinions are not unanimous. While most of those who hold that the whole heaven is finite say that the earth lies at the centre, the philosophers of Italy, the so-called Pythagoreans, assert the contrary. They say that in the middle there is fire, and that the earth is one of the stars, and by its circular motion round the centre produces night and day. They also construct another earth opposite to ours, which they call counter-earth, and in this they are not seeking explanations and causes to fit the observed phenomena, but they are rather straining the phenomena in the effort to make them agree with certain explanations and views of their own. Many others might agree with them that the place in the centre should not be assigned to the earth, if they looked for confirmation, not to the observed facts, but to *a priori* arguments. For they consider that the worthiest place is appropriate to the worthiest occupant, and fire is worthier than earth, and the limit worthier than the intervening parts, while the extremity and the centre are limits; arguing from these considerations, they think that it is not the earth which lies in the centre of the (heavenly) sphere, but rather the fire. Further, the Pythagoreans give the additional reason that it is the most sovereign part of the All which ought to be most safeguarded, and that is the centre; they accordingly give this, or the fire which occupies this place, the name of Zeus' Watch-tower, implying that the word "centre" is unambiguous, and that the centre of a magnitude is also the centre of the corresponding *thing*, or its natural centre. . . . Such, then, are the opinions of certain persons about the position of the earth. Similarly, with regard to its rest or motion there is not universal agreement. Those who say that the earth does not so much as occupy the centre make it revolve in a circle round the centre, and not only the earth, but the counter-earth also, as we said before. Some, again, think that there may be even more bodies of the kind revolving round the centre; only they are invisible to us because of the interposition of the earth. This they give as the reason why there are more eclipses of the moon than of the sun; for each of the revolving bodies, and not only the earth, may obscure the moon. The fact that the earth is not the centre, but at a distance represented by the whole hemisphere, constitutes, in their opinion, no reason why the phenomena should not present the same appearance to us if we lived away from the centre, as they would on the assumption that the earth was in the centre; seeing that, as it is, there is nothing to indicate that we are at a distance (from the centre) represented by half the earth's diameter.

Simplicius, on *De caelo*, pp. 511, 25–512, 1; 512, 9–17, Heib.

The Pythagoreans, on the other hand, say that the earth is not at the centre, but that in the centre of the universe is fire, while round the centre revolves the counter-earth, itself an earth and called counter-earth because it is opposite to our earth; then next to the counter-earth comes our earth, which itself also revolves about the centre, and next to the earth the moon; this is stated by Aristotle in his work on the Pythagoreans. The earth, then, being like one of the stars, moves round the centre and, according to its position with reference to the sun, makes night and day. The counter-earth, as it moves round the centre and accompanies our earth, is invisible to us, because the body of the earth is continually interposed in our way. . . . The more genuine exponents of the doctrine describe as fire at the centre the creative force which from the centre imparts life to all the earth, and warms afresh the part of it which has cooled. Hence some call this fire the Tower of Zeus, as Aristotle states in his Pythagorean Philosophy, others the Watch-tower of Zeus, as Aristotle calls it here, and others, again, the Throne of Zeus, if we may credit different authorities. They called the earth a star as being itself, too, an instrument of time, for it is the cause of days and nights, since it makes day when it is lit up in that part of it which faces the sun, and it makes night throughout the cone formed by its shadow.

Aëtius, II, 7, 7.

Philolaus calls the fire in the middle about the centre the Hearth of the Universe, the House of Zeus, the Mother of the Gods, the Altar, Bond and Measure of Nature. And again he assumes another fire in the uppermost place, the fire which encloses (all). Now the middle is naturally first in order, and round it ten divine bodies move as in a dance, [the heaven and] <after the sphere of the fixed stars> the five planets, after them the sun, under it the moon, under the moon the earth, and under the earth the counter-earth; after all these comes the fire which is placed like a hearth round the centre. The uppermost part of that which encloses (all), wherein the elements exist in all their purity, he calls Olympus, and the parts under the moving Olympus, where are ranged the five planets with the moon and the sun, he calls the Universe, and lastly, the part below these, the part below the moon and round the earth, where are the things which suffer change and becoming, he calls the Heaven.

Ib., III, 11, 3.

Philolaus the Pythagorean places the fire in the middle, for this is the Hearth of the All; second to it he puts the counter-earth, and third the inhabited earth, which is placed opposite to, and revolves with, the counter-earth; this is the reason why those who live in the counter-earth are invisible to those who live in our earth.

Ib., II, 4, 15.

The governing principle is placed in the fire at the very centre, and the Creating God established it there as a sort of keel to the (sphere) of the All.

Ib., III, 13, 1, 2.

Others maintain that the earth remains at rest. But Philolaus the Pythagorean held that it revolves round the fire in an oblique circle, in the same way as the sun and moon.

The "harmony of the spheres"

ARISTOTLE, *Metaph.* A 5, 986 a 1.

They (the Pythagoreans) conceived that the whole heaven is harmony and number; thus, whatever admitted facts they were in a position to prove in the domain of numbers and harmonies, they put these together and adapted them to the properties and parts of the heaven and its whole arrangement. And if there was anything wanting anywhere, they left no stone unturned to make their whole system coherent. For example, regarding as they do the number ten as perfect and as embracing the whole nature of numbers, they say that the bodies moving in the heaven are also ten in number, and, as those which we see are only nine, they make the counter-earth the tenth.

HIPPOLYTUS, *Refut.* I, 2, 2.

Pythagoras maintained that the universe *sings*, and is constructed in accordance with a harmony; and he was the first to reduce the motion of the seven heavenly bodies to rhythm and song.

ALEXANDER, in *Metaph.* A 5, p. 542 a 5–18 Brandis.

They (the Pythagoreans) said that the bodies which revolve round the centre have their distances in proportion, and some revolve more quickly, others more slowly, the sound which they make during this motion being deep in the case of the slower, and high in the case of the quicker; these sounds then, depending on the ratio of the distances, are such that their combined effect is harmonious. . . . Thus, the distance of the sun from the earth being, say, double the distance of the moon, that of Aphrodite triple, and that of Hermes quadruple, they considered that there was some arithmetical ratio in the case of the other planets as well, and that the movement of the heaven is harmonious. They said that those bodies move most quickly which move at the greatest distance, that those bodies move most slowly which are at the least distance, and that the bodies at intermediate distances move at speeds corresponding to the sizes of their orbits.

The sun

Aëtius, II, 22, 5.

The Pythagoreans held the sun to be spherical.

Ib., II, 20, 12.

Philolaus the Pythagorean holds that the sun is transparent like glass, and that it receives the reflection of the fire in the universe, and transmits to us both light and warmth, so that there are, in some sort, two suns: the fiery (substance) in the heaven, and the fiery (emanation) from it which is mirrored, as it were, not to speak of a third also, namely, the beams that are scattered in our direction from the mirror by way of reflection (or refraction); for we give this third also the name of sun, which is thus, as it were, an image of an image.

LEUCIPPUS

Diogenes Laërtius, IX, 31 sqq.

Leucippus says that the All is infinite, and that it is part full, and part empty. These (the full and the empty), he says, are the elements. From them arise innumerable worlds, and are resolved into them. The worlds come into being thus. There were borne along by "abscission from the infinite" many bodies of all sorts of figures "into a mighty void," and they being gathered together produce a single vortex. In it, as they came into collision with one another and were whirled round in all manner of ways, those which were alike were separated apart and came to their likes. But, as they were no longer able to revolve in equilibrium owing to their multitude, those of them that were fine went out to the external void, as if passed through a sieve; the rest stayed together, and becoming entangled with one another, ran down together, and made a first spherical structure. This was in substance like a membrane or skin containing in itself all kinds of bodies. And, as these bodies were borne round in a vortex, in virtue of the resistance of the middle, the surrounding membrane became thin, as the contiguous bodies kept flowing together from contact with the vortex. And in this way the earth came into being, those things which had been borne towards the middle abiding there. Moreover, the containing membrane was increased by the further separating out of bodies from outside; and, being itself carried round in a vortex, it further got possession of all with which it had come in contact. Some of these becoming entangled, produced a structure, which was at first moist and muddy; but, when they had been dried and were revolving along with the vortex of the whole, they were then ignited and produced the substance of the heavenly bodies. The circle of the sun is the outermost, that of the moon is nearest to the earth, and those of the others are between these. And all the heavenly bodies are ignited because of the swiftness of their motion; while the sun is also ignited by the stars. But the moon only receives a small portion of fire. The sun and the moon are eclipsed. . . . (And the obliquity of the zodiac is produced) by the earth being inclined towards the south; and the northern parts of it have constant snow and are cold and frozen. And the sun is eclipsed rarely, and the moon continually, because their circles are unequal. And, just as there are comings into being of the world, so there are growths and decays and passings away in virtue of a certain necessity, of the nature of which he gives no clear account.

John Burnet.

Ib., IX, 30.

The worlds arose through bodies falling into the void and becoming entangled with one another; from the motion, as the bodies increased, the stars arose. The sun is carried in a larger circle about the moon; the earth rides through being whirled round about the middle; its shape is that of a tambourine.

DEMOCRITUS

HIPPOLYTUS, *Refut.* I, 13.

DEMOCRITUS said that there are worlds infinite in number and differing in size. In some there is neither sun nor moon, in others the sun and moon are greater than with us, in others there are more than one sun and moon. The distances between the worlds are unequal, in some directions there are more of them, in some, fewer; some are growing, others are at their prime, and others again declining, in one direction they are coming into being, in another they are ceasing to be. Their destruction comes about through collision with one another. Some worlds are destitute of animal and plant life and of all moisture. In our world, the earth came into being before the stars; the moon has the lowest place, then the sun, <then the other planets,> and after them the fixed stars. The planets themselves even are not at equal heights (i.e. distances). A world continues to be in its prime only until it becomes incapable of taking to it anything from without.

AËTIUS, II, 13, 20, 25.

(13, 4) Democritus says the stars are stones.

(20, 7) The sun is a red-hot mass, or a stone on fire.

(25, 9) (The moon has the appearance of being earthy) because it shows a sort of shadow of lofty elevations in it; it has hollows or valleys.

Ibid., III, 10, 5.

Democritus held that the earth is disk-like laterally, but hollowed out in the middle.

Agathemerus, I, 1, 2.

The ancients described the inhabited earth as round, and regarded Greece as lying in the middle of it, and Delphi as the centre of Greece. But Democritus, a much-travelled man, said that the (inhabited) earth was elongated, its length being one and a half times its breadth.

Aëtius, III, 12, 2.

Democritus said that the earth as it grew became inclined southwards, because the southern portion of that which enveloped it was weaker; for the northern regions are intemperate (i.e. frigid), while the southern are temperate; hence it is in the south that the earth *sags*, namely, where fruits and all growths are in excess.

Lucretius, V, 621–636.

It seems highly probable that that is true which the revered judgment of the great Democritus lays down, that, the nearer each of the stars is to the earth, the less it can be carried round in the revolution of the heaven. For the swift, whirling force of the heaven diminishes lower down, and therefore the sun is gradually left behind with the rearward signs because it is much lower than the burning signs. And still more the moon; the lower and farther from the heaven her course is, and the nearer she is to the earth, the less can she keep pace with the signs. For the weaker the whirl in which, being below the sun, she is carried, the more all the signs about her catch her up and pass her. Hence it is that the moon seems to return to each sign more promptly, because the signs revert to her more quickly.

PLATO

The study of astronomy

Epinomis, 990 A, B.

THE true astronomer must be the wisest of men, understanding by the term, not the man who cultivates astronomy in the manner of Hesiod and all others of that type, concerning himself only with such things as risings and settings, but the man who investigates the seven revolutions included in the eight revolutions [i.e. the revolutions of the sun, the moon, and the five planets, apart from the eighth motion, that of the daily rotation], each of the seven describing its own circle in a manner such as would never be easily comprehended by any one unless he possessed extraordinary powers.

Republic, VII, 529 A–530 B.

"You seem with sublime self-confidence to have formed your own conception of the nature of the learning which deals with the things above. At that rate, if a person were to throw his head back and learn something by contemplating a carved ceiling, you would probably suppose him to be investigating it, not with his eyes, but with his mind. You may be right, and I may be wrong. But I, for my part, cannot think any other study to be one that makes the soul look upwards except that which is concerned with the real and the invisible, and, if any one attempts to learn anything that is *perceivable*, I do not care whether he looks upwards with mouth gaping or downwards with mouth shut; he will never, as I hold, learn—because no object of sense admits of knowledge—and I maintain that in that case he is not looking upwards but downwards, even though the learner float face upwards on land or in the sea." "I stand corrected," said he; "your rebuke was just. But what is the way, different from the present method, in which astronomy should be studied for the purposes we have in view?"

"This," said I, "is what I mean. Yonder broideries in the heaven, forasmuch as they are broidered on a visible ground, are properly considered to be more beautiful and perfect than anything else that is visible; yet they are far inferior to those which are true, far inferior to the movements wherewith essential speed and essential slowness, in true number and in all true forms, move in relation to one another and cause that which is essential in them to move: the true objects which are apprehended by reason and intelligence, not by sight. Or do you think otherwise?" "Not at all," said he. "Then," said I, "we should use the broideries in the heaven as illustrations to facilitate the study which aims at those higher objects, just as we might employ, if we met with them, diagrams drawn and elaborated with exceptional skill by Daedalus or any other artist or draughtsman; for I take it that any one acquainted with geometry who saw such diagrams would indeed think them most beautifully finished, but would regard it as ridiculous to study them seriously in the hope of gathering from them true relations of equality, doubleness, or any other ratio." "Yes, of course, it would be ridiculous," he said. "Then," said I, "do you not suppose that one who is a true astronomer will have the same feeling when he looks at the movements of the stars? That is, will he not regard the maker of the heavens as having constructed them and all that is in them with the utmost beauty of which such works admit; yet, in the matter of the proportion which the night bears to the day, both of these to the month, the month to the year, and the other stars to the sun and moon and to one another, will he not, think you, regard as absurd the man who supposes these things, which are corporeal and visible, to be changeless and subject to no aberrations of any kind; and will he not hold it absurd to exhaust every possible effort to apprehend their true condition?" "Yes, I for my part certainly think so, now that I hear you state it." "Hence," said I, "we shall pursue astronomy, as we do geometry, by means of problems, and we shall dispense with the starry heavens, if we propose to obtain a real knowledge of astronomy, and by that means to convert the natural intelligence of the soul from a useless to a useful possession." "The plan which you prescribe is certainly far more laborious than the present mode of studying astronomy."

The Heavenly Choir

Phaedrus, 246 E–247 C.

Zeus, the great captain in heaven, mounted on his winged chariot, goes first and disposes and oversees all things. Him follows the army of gods and daemons ordered in eleven divisions; for Hestia alone abides in the House of God, while, among the other gods, those who are of the number of the twelve, and are appointed to command, lead the divisions to which they were severally appointed.

Many glorious sights there are of the courses in the heavens traversed by the race of blessed gods, as each goes about his own business; and whosoever wills, and is able, follows, for envy has no place among the Heavenly Choir. . . .

The chariots of the gods move evenly, and, being always obedient to the hand of the charioteer, travel easily; the others travel with great difficulty.

The souls, which are called immortal, when they are come to the summit of the Heaven, go outside and stand on the roof, and, as they stand, they are carried round by its revolution and behold the things that are outside the Heaven.

Anaxagoras and Mind

Phaedo, 97 B–99 B.

When once I heard someone reading from a book, as he said, of Anaxagoras, in which the author asserts that it is Mind which disposes and causes all things, I was pleased with this cause, as it seemed to me right in a certain way that Mind should be the cause of all things, and I thought that, if this is so, and Mind disposes everything, it must place each thing as is best. If then any one wished, in the case of anything, to discover the cause of its coming to be, being destroyed, or being, in the way it is, what he has to find out about it is how it is best for it to be or to act or be acted upon in any other way whatever. . . . With these considerations in view, I was glad to think that I had found a guide entirely to my mind in this matter of the cause of existing things, I mean Anaxagoras, and that he would first tell me whether the earth is flat or round and, when he had told me this, would add to it an explanation of the cause and the necessity for it, which would be the Better, that is to say, that it is better that the earth should be as it is; and further, if he should assert that it is in the centre, that he would add the explanation that it was better that it should be in the centre. If he should assert this, I was prepared not to press for any other sort of cause. Similarly I was prepared to be told in like manner, with regard to the sun, the moon, and the other stars, their relative speeds, their turnings-back, and their other conditions, in what way it is best for each of them to be, to act, or to be acted upon so far as they are acted upon. For I should never have supposed that, when once he had said that these things were ordered by Mind, he would have assigned to them in addition any cause except the fact that it is best that they should be as they are. . . . From what a height of hope, then, was I hurled down when I went on with my reading and saw a man that made no use of Mind for ordering things, but assigned, as their cause, airs, aethers, waters, and any number of other absurdities. He seemed to me to be, for all the world, in the same position as a man who, after saying that, whatever Socrates does he does by Mind, should then try to give the causes of everything I do, and say first that the reason of my being seated here is that the body is composed of bones and sinews, that the bones are solid, and have joints separating them one from another, and that the sinews are capable of contraction and relaxation, while they surround the bones along with flesh and skin which keep them together; the bones then being lifted in their own sockets, the sinews, by relaxing and contracting, make me somehow able at a particular moment to bend my limbs, and this is the cause of my being huddled up and being seated here. . . . Not to be able to discern that the cause for what is is one thing, but that that in the absence of which the cause could never be a cause is another thing, suggests that most people are fumbling in the dark and, when they call the latter the cause, are using the wrong name. Thus it is that one purports to make the earth remain stationary by enclosing it in a vortex, while another places the air as a support to the earth which he regards as like a flat kneading-trough.

The earth

Phaedo, 108 C–110 A.

There are many and wondrous regions in the earth, and it is neither in its nature nor in its size what it is supposed to be by those whom we commonly hear speak about it; of this I have been convinced, I will not say by whom. . . . My persuasion as to the form of the earth and the regions within it I need not hesitate to tell you. . . . I am convinced, then, said he, that in the first place, if the earth, being a sphere, is in the middle of the heaven, it has no need either of air or of any other such force to keep it from falling, but that the uniformity of the substance of the heaven in all its parts, and the equilibrium of the earth itself, suffice to hold it; for a thing in equilibrium in the middle of any uniform substance will not have cause to incline more or less in any direction, but will remain as it is, without any such inclination. In the first place, I am persuaded of this.

Moreover, I am convinced that the earth is very

great, and that we who live in the region from the river Phasis to the Pillars of Heracles inhabit a small part of it; like to ants or frogs round a pool, so we dwell round the sea; while there are many other men dwelling elsewhere in many regions of the same kind. For everywhere on the earth's surface there are many hollows of all kinds both as regards shapes and sizes, into which water, clouds, and air flow and are gathered together; but the earth itself abides pure in the purity of the heaven in which are the stars, the heaven, which the most part of those who use to speak about these things call aether, and it is the sediment of the aether which, in the forms we mentioned, is always flowing and being gathered together in the hollow places of the earth. We, then, dwelling in the hollow parts of it, are not aware of the fact, but imagine that we dwell above on the surface; this is just as if any one dwelling down at the bottom of the sea were to imagine that he dwelt on its surface, and, beholding the sun and the other heavenly bodies through the water, were to suppose the sea to be the heaven, for the reason that, through being sluggish and weak, he had never yet risen to the top of the sea, nor been able, by putting forth his head and coming up out of the sea into the place where we live, to see how much purer and more beautiful it is than his abode, neither had heard this from another who had seen it. We are in the same case; for, though dwelling in a hollow of the earth, we think we dwell on its surface, and we call the air heaven as though this were the heaven, and through this the stars moved, whereas, in fact, we are through weakness and sluggishness unable to pass through and reach the limit of the air; for, if any one could reach the top of it, or could get wings and fly up, then, just as fishes here, when they come up out of the sea, espy the things here, so he, having come up, would likewise descry the things there, and, if his strength could endure the sight, would know that there is the true heaven, the true light, and the true earth. For here the earth, with its stones and the whole place where we are, is corrupted and eaten away, as things in the sea are eaten away by the salt, insomuch that there grows in the sea nothing of moment nor anything perfect, so to speak, but there are hollow rocks, sand, clay without end, and sloughs of mire wherever there is also earth, things not worthy at all to be compared to the beautiful objects within our view; but the things beyond would appear to surpass even more the things here.

The Myth of Er

Republic, X, 616 B–617 D.

Now when seven days had passed since the spirits arrived in the meadow, they were compelled to arise on the eighth day and journey thence; and on the fourth day they arrived at a point from which they saw extended from above, through the whole heaven and earth, a straight light, like a pillar, most like to the rainbow, but brighter and purer. This light they reached when they had gone forward a day's journey, and there, at the middle of the light, they saw, extended from heaven, the extremities of the chains thereof; for this light it is that binds the heaven together, holding together the whole revolving firmament as the undergirths hold together triremes; and from the extremities they saw extended the Spindle of Necessity by which all the revolutions are kept up.

The shaft and hook thereof are made of adamant, and the whorl is partly of adamant and partly of other substances.

Now the whorl is after this fashion. Its shape is like that we use; but from what he said we must conceive of it as if, in one great whorl, hollow and scooped out through and through, there were inserted another whorl of the same kind but smaller, nicely fitting into it, like those boxes which fit into one another; and into this again we must suppose a third whorl fitted, into this a fourth, and after that four more. For the whorls are altogether eight in number, set one within another, showing their rims above as circles and forming about the shaft a continuous surface as of one whorl; while the shaft is driven right through the middle of the eighth whorl.

The first and outermost whorl has the circle of its rim the broadest, that of the sixth is second in breadth, that of the fourth is third, that of the eighth is fourth, that of the seventh is fifth, that of the fifth is sixth, that of the third is seventh, and that of the second is eighth. And the circle of the greatest is of many colours, that of the seventh is brightest, that of the eighth has its colour from the seventh which shines upon it, that of the second and fifth are like each other, and yellower than those aforesaid, the third is the whitest in colour, the fourth is pale red, and the sixth is the second in whiteness.

The Spindle turns round as a whole with one motion, and within the whole, as it revolves, the seven inner circles revolve slowly in the opposite sense to the whole, and of these the eighth goes the most swiftly; second in speed and all together go the seventh and sixth and fifth; third in the speed of its counter-revolution the fourth appears to move, fourth in speed comes the third, and fifth the second. And the whole Spindle turns in the lap of Necessity.

Upon each of the circles above stands a Siren, carried round with it, and uttering one single sound, one single note, and out of all the notes, eight in number, is formed one harmony.

And again, round about, sit three others at equal distances apart, each on a throne, the daughters of Necessity, the Fates, clothed in white raiment and with garlands on their heads, Lachesis, Clotho, and Atropos, and they chant to the harmony of the Sirens, Lachesis, the things that have been; Clotho, the things that are; and Atropos, the things that shall be.

And Clotho at intervals, with her right hand, takes hold of the outer revolving whorl of the Spindle and helps to turn it; Atropos, with her left hand, does the same to the inner whorls; Lachesis, with both hands, takes hold of the outer and inner alternately (i.e. of the outer with her right hand, and of the inner with her left).

The creation of the universe

Timaeus, 33 B–34 B.

And he gave the universe the figure which is proper and natural. For the living thing which should contain within itself all living things that figure would be proper which contains within itself all figures whatsoever. Wherefore he turned it, as in a lathe, round and spherical, with its extremities equidistant in all directions from the centre, the figure of all figures most perfect and most like to itself, for he deemed the like more beautiful than the unlike. To the whole he gave, on the outside round about, a surface perfectly finished and smooth, for many reasons. It had no need of eyes, for nothing visible was left outside it; nor of hearing, for there was nothing audible outside it; and there was no breath outside it requiring to be inhaled. Nor again did it need to have any organ whereby to take to itself sustenance and, when digested, to void it again; for nothing went forth from it nor came to it from anywhere, since nothing existed (but itself). It was designedly created such as to provide itself with its own sustenance out of its own waste, and to act and be acted upon wholly in itself and by itself; for its Creator deemed that, were it self-sufficing, it would be much better than if it had need of other things. He did not think he ought uselessly to give it hands, which it could not need for taking hold of anything or for defending itself against anything; nor yet feet, nor generally anything that serves for taking steps. For he allotted to it the motion which was proper to its bodily form, that motion of the seven motions which is most bound up with understanding and intelligence. Wherefore, turning it round in one and the same place upon itself, he made it move with circular rotation; all the other six motions he took away from it and made it exempt from their wanderings. And since for this revolution it had no need of feet, he created it without legs and without feet.

This, then, was the whole reasoned purpose of the God who forever is for the God who was to be; in pursuance of this purpose he made it smooth and even and everywhere equidistant from the centre, a body whole and perfect, made up of perfect bodies. And God set soul in the midst of it and spread it throughout, and wrapped the body about it from outside, and established it as a circle (sphere) with circular rotation, one heaven, single and alone, able through its excellence to commune with itself, needing no other companion, but sufficient to itself as companion and friend. For all these reasons he created it a blessed God.

Ib., 34 c–36 D

God made soul in birth and excellence prior to and older than body to be its mistress and governor; and he framed it out of the following elements and in the following way. . . .

[Here follows the hopelessly difficult passage, 35 A–36 c, describing the creation of the soul out of a mixture of the constituents Same, Other, and Essence, divided up in accordance with the intervals of a musical scale, so that harmony pervaded the whole substance. The interpretation of this passage cannot be entered upon here. Having arrived at a "soul-strip" as it were, Plato proceeds to deal with it as follows, 36 C–D:]

Next he cleft the structure so formed lengthwise into two halves and, laying them across one another, middle upon middle, in the shape of the letter *Chi* (χ), he bent them in a circle and joined them, making them meet themselves and each other at a point opposite to that of their original contact; and he comprehended them in that motion which revolves uniformly and in the same place, and one of the circles he made exterior and one interior. The exterior movement he named the movement of the Same; the interior the movement of the Other. The revolution of the circle of the Same he made to follow the side (of a rectangle) towards the right hand, that of the circle of the Other he made to follow the diagonal and towards the left hand, and he gave the mastery to the revolution of the Same and uniform, for he left that single and undivided; but the inner circle he cleft, by six divisions, into seven unequal circles in the proportion severally of the double and triple intervals, each being three in number; and he appointed that the circles should move in opposite senses, three at the same speed, and the other four differing in speed from the three and among themselves, yet moving in a due ratio.

Time : sun, moon, and planets

Timaeus, 37 D–E, 38 C–39 D.

[The Creator's pattern, having been a living being externally existent, and it being impossible that a created thing should be eternal,] God bethought him to make a moving image of eternity and, while he was ordering the heaven, he made an eternal image, proceeding according to number, of eternity that abides in unity, this image being what we have named Time. For, whereas days and nights and months and years were not before the heaven was created, he then contrived their birth along with the construction of the heaven. Now these are all portions of time. . . .

This, then, being the plan and intent of God for the birth of time, that time might be generated, the sun, the moon and the five other stars which are called planets were made for defining and preserving the numbers of time. And when God had made their several bodies, he set them in the orbits in which the revolution of the Other was proceeding, in seven orbits seven stars. The moon he placed in that nearest the earth; in the second above the earth he placed the sun; next, the Morning Star and that which is held sacred to Hermes he placed in those (orbits) which proceed in a circle having equal speed with the sun, but have the contrary tendency to it; hence it is that the sun and the star of Hermes and the Morning Star overtake, and are in like manner overtaken by, one another. And as to the rest, if we were to set forth the orbits in which he placed them, and the causes for which he did so, the account, though only by the way, would lay on us a heavier task than that which is our main purpose. These things may perhaps hereafter receive proper treatment, when we have leisure.

But when each of the beings [the sun, moon, and planets] which were to join in the making of time had arrived in its proper orbit, and they had been as animate bodies secured with living bonds and had learnt their appointed task, then in the motion of the Other, which was oblique and crossed the motion of the Same and was controlled by it, one planet described a larger, and another a smaller circle, and those which described the smaller went round it more swiftly, and those which described the larger more slowly; but by reason of the motion of the Same, those which went round more swiftly appeared to be overtaken by those which went round more slowly, though in reality they overtook them. For the motion of the Same, which twists all their circles into spirals, because they have two separate and simultaneous motions in opposite senses, is the swiftest of all, and makes that which departs most slowly from it appear closest to itself.

And, that there might be some visible measure of their relative slowness and swiftness, and that the eight revolutions might proceed, God kindled a light in the second orbit from the earth, which we have named the Sun, in order that it might shine most brightly throughout the whole heaven, and that living things, so many as was meet, should possess number, learning it from the revolution of the Same and uniform. Night, then, and day have come to be in this way and for these reasons, (making up) the period of the one and most intelligent revolution; a month has passed when the moon, having completed her own orbit, overtakes the sun, and a year when the sun has completed its own orbit.

But the courses of the others men have not grasped, save a few out of many; and they neither give them names nor investigate the measurement of them one against another by numerical calculation, in fact, they can scarcely be said to know that time is represented by their wanderings, which are incalculable in multitude and marvellously intricate.

None the less, however, can we observe that the perfect number of time fulfils the perfect year at the moment when the relative speeds of all the eight revolutions accomplish their course together and reach their starting-point, being measured by the circle of the Same and uniformly moving.

Form and movements of fixed stars

Timaeus, 40 A–B.

The visible form of the divine he made chiefly of fire, that it might be most bright and most beautiful to behold, and, likening it to the All, he fashioned it like a sphere, and set it in the mind of the supreme to follow after it; and he disposed it round about throughout the whole heaven, to be an adornment of it in very truth, broidered over the whole expanse. And he attached two movements to each, one in the same place and uniform, as ever thinking the same thoughts about the same things, the other a movement forward mastered by the revolution of the Same and uniform; but for the other five movements he made it motionless and at rest, in order that each star might attain the highest order of perfection.

From this cause, then, came to be all the stars that wander not, living beings, divine, eternal, which abide for ever revolving uniformly and in the same place; while those that turn back and wander as aforesaid came to be in the manner already described.

The earth and the planets

Timaeus, 40 B–D.

But the earth our foster-mother, rolling in its course about the axis stretched through the universe, he contrived for a guardian and maker of night and day, the first and chiefest of the gods that have come into being within the heaven.

But the dances of these same gods, their comings alongside one another and the returnings of their orbits upon themselves, and their advancings, and the questions which of the deities meet one another in their conjunctions and which are in opposition, in what order they pass before one another, and at what times they are hidden from us, and again reappearing send, to them who cannot calculate their movements terrors, and portents of things to come—to declare all this without visible imitations of these movements were labour lost.

Laws, VII, 821 B–822 C.

Clinias. You are probably right; but what lesson about the stars shall we find that has this character [that of being beautiful and true, advantageous to the state, and pleasing to God]?

Athenian Stranger. My good friends, I make bold to say that nowadays we Greeks all say what is untrue about the great gods, the sun and the moon alike.

Cl. What is the falsehood you mean?

Ath. We say that they never keep to the same path, and we say the same of certain other stars as well, calling them planets.

Cl. By Zeus, O stranger, you are right. I myself have many times in my life noticed that the Morning Star and the Evening Star and some others never keep to the same course, but wander in every possible way, and, of course, the sun and moon behave in the way that is familiar to us all.

Ath. These are just the things, Megillus and Clinias, which citizens of our country at all events, and the young, ought to learn about the gods in heaven; they ought to learn the facts about all of them so far as not only to avoid slandering them in this respect, but to honour them at all times, sacrificing to them, and addressing to them pious prayers.

Cl. You are right, assuming that it is first possible to understand what you are driving at; then, if there is anything in what we now say about the gods that is not correct, and instruction will correct it, I, too, agree that we ought to learn a thing of such magnitude and importance. Therefore, by all means try to explain how these things are as you say, and we will try to follow what you tell us.

Ath. Well, it is not easy to take in what I mean, nor yet is it very difficult or a very long business; witness the fact that, although it is not a thing which I learnt when I was young or very long ago, I can now, without taking much time, make it known to you; whereas, if it had been difficult, I, at my age, should never have been able to explain it to you at yours.

Cl. I dare say. But what sort of lesson is this you speak of, which you call surprising and proper to be taught to the young, but which we do not know? Try to tell us this much about it as clearly as you can.

Ath. I will try. Well, my good friends, this view which is held about the moon, the sun, and the other stars, to the effect that they ever wander, is not correct, but the fact is the very contrary of this. For each of them traverses the same path, not many paths, but always one, in a circle, whereas it appears to move in many paths. And, again, the swiftest of them is incorrectly thought to be the slowest, and vice versa. Now, if the truth is one way, and we think another way, it is as if we had the same idea about racing horses, or long-distance runners at Olympia, and were to address the swiftest as the slowest, and the slowest as the swiftest, and to award the praise accordingly, knowing all the time that the so-called loser had really won—in which case, I imagine that we should not be awarding the praise in the proper way or a way agreeable to the runners, who are only human. Seeing, then, that we now make the very same mistake about the gods, should we not expect that what would have been ridiculous and wrong in the case of the races would, here and in this question, be—well, by no means a laughing matter; nay, it would not even be consistent with respect for the gods if we repeated a false report against them.

Epinomis, 982 c–983 c.

Now to prove that the stars, with all this journeying, have intelligence, men should have found sufficient evidence in the fact that the stars do the same things always, because they have for an unimaginable length of time been doing things determined on from of old, and they do not, by changing their intent this way and that, and doing one thing at one time and another at another, come to wander and change their orbits. Most of us, it is true, have thought the very contrary, namely that, because they are always doing the same things in the same way, they have no soul: in this the generality of people follow the lead of the unintelligent, their idea being that men are intelligent and alive because they move, but that the divine is unintelligent because it abides always in the same courses: yet a man holding to what is fairer and better and pleasing should have been able to assume that we ought, for that very reason, to think that that is intelligent which acts always in the same place, in the same manner, and for the same reasons, and that this is the natural constitution of the stars, which not only are most fair to view, but, by moving in a procession and a dance the most beautiful and impressive of all dances, minister to the needs of all living things. Further, to show that we are right in calling them animate beings, let us first consider their size. For they are not really the small things they appear to be, but each of them is of incredible mass—this deserves to be believed because the assumption has been sufficiently proved—thus the sun as a whole may properly be conceived to be greater than the earth as a whole, and all the moving stars are of stupendous size. Let us then think how it would be possible for a natural force of any kind whatever to carry round a mass so huge as we actually see carried round, and in a time which is always the same. I say that it must be God who is the cause; otherwise it could never happen; for nothing can ever become animate except by reason of God, as we have declared. When God has this power, it is to him a supremely easy matter that any particular body, and any whole mass, should first become a living being, and then that he should carry it in whatever direction he may conceive to be best. Now, therefore, we can propound one statement which is true about all these things: it is not possible that the earth and the heaven, the stars, and the masses as a whole which they comprise should, if they have no soul attached to each body or dwelling in each body, nevertheless accurately describe their orbits in the way they do, year by year, month by month, and day by day, and that all of us should receive all the blessings which actually come to us.

The stars animate beings: motions and names of planets

Epinomis, 986 A–988 E.

Know that of the powers created in the whole heaven eight are akin to one another; these I have observed—and this is no great accomplishment of mine, for any one else can easily see it. Three of these are the following:

one is that of the sun, one that of the moon, and one that of the stars which we mentioned a short time ago; and there are five others. And with regard to all these and those bodies in them which travel either of their own motion, or because they are borne on vehicles and so travel, let none among all of us ever suppose that some of them are gods and others not, nor that some are legitimate, while others are such as it would not be lawful for any one of us even to name, but let us all say and affirm that they are all in a brotherly relation and share as brothers, and let us worship them without assigning to one a year, to another a month, and to others a certain share or a certain time, namely that in which each of them traverses its own circuit, and so helps to complete the order of the universe which the reason most divine of all ordained to be visible: a work which he who is blessed first marvelled at, and then felt a passion to understand as much of it as is within the power of mortals, deeming that he will in this way pass the best and most happy life, after death will pass to regions appropriate to such excellence, and then, after a true and real initiation, having attained to the share which one man may have in the wisdom which is one, and having become a spectator for the rest of time of the fairest sights, will so continue.

Now, for the rest, it remains for us to say (of the gods in question) how many they are, and which; heaven forbid that we should ever be convicted of falsehood. This much then I firmly maintain. I say again that they are eight, and of the eight three have been mentioned, while five remain. The fourth orbit or course, together with the fifth, is about equal in speed to that of the sun and, in general, neither slower nor swifter. These being three in number, they must be so considered by any one with adequate intelligence. Let us then say that these movements are those of the sun, the Morning Star, and a third. Of the third we cannot speak by name, because its name is not known, and the reason of this is that the first man who observed these things was a barbarian. An ancient custom saw to the maintenance of those who were the first to notice these facts, as they were able to do owing to the beauty of the summer season, enjoyed in a sufficient degree by both Egypt and Syria; to observers in those countries the stars are, at all times, all visible together roughly speaking, because they live always far removed from the clouds and waters in the universe; and it is from thence that the results of their observations reach all other quarters as well as our country, after having been tested through untold, nay, endless ages. Hence we should not hesitate to provide for these things in our laws; for to rank things divine as not worthy of honour, while regarding other things as worthy, is clearly unintelligent. As for the want of names for them, we must hold this (latter) attitude to be responsible. In fact, however, they have been called after gods. The Morning and Evening Star, which are one and the same, belong to Aphrodite, and this is fairly reasonable, the name being a very proper one for a Syrian lawgiver to assign to it; the star, however, which in its course accompanies both the sun and the star aforesaid is generally called that of Hermes. Let us now speak of the three further revolutions the course of which is to the right, together with those of the sun and moon.—One (other) we must describe as the eighth, which one would best call the universe, and which travels in a sense contrary to all the others, <not>

carrying the others with it as might be supposed by men with little knowledge of these things.[1] But what we adequately know we are bound to assert and do assert; for that which is the true wisdom is in this sort of way revealed to the man who has even a small share in right, that is, divine, understanding.—There remain, then, three stars, one of which is distinguished from the others by its slowness, and some call it by the name of Kronos; the next to it in slowness we must call the star of Zeus, and that next to it is the star of Ares, and this last has the reddest colour of them all. None of these things is difficult for any one to understand, if explained by someone; but when we have learnt them we must, as we maintain, believe them.

One thing every Greek ought to bear in mind, and that is that the situation of our race is about the most favourable of all for excellence. What is commendable in it we must hold to be that it may be described as midway between the extremes of winter storms and summer

[1] The intention here is apparently to assert the contrary of the view taken in the *Timaeus*. In the *Timaeus* the motion of the circle of the Same "controls" or "prevails over" that of the circle of the Other; in other words, the daily rotation carries the sun, moon, and planets (no less than the fixed stars) with it, though they have, in addition, their own motions in the opposite sense along the zodiac circle. What it is here intended to convey is apparently that the daily rotation does *not* carry the other revolutions with it, and that it is only an ignorant person who could think that it does. Burnet thought to make this clearer by inserting "not" (οὐκ) before "carrying"; but the odd thing is that we can get exactly the same meaning out of the sentence without the "not"; it simply depends on the way you read it out; it is a question of intonation simply.

heat. It is our disadvantage in relation to summer weather as compared with those other (eastern) localities which is responsible for the observations of these gods in the universe reaching the Greeks later, as we said. Let us, however, take it for granted that, whatever the Greeks take over from the barbarians, they elaborate it to a finer perfection. So then, in the matters of which we are now speaking, we must reflect that it is difficult to discover all such things with absolute certainty, but there is a great and withal bright prospect that the Greeks will worship all these gods in a manner truly more beautiful and more just than is the report and service of them which have passed to us from the barbarians, seeing that we have the benefit of education and of oracles from Delphi, as well as of all the service provided for in our laws. And let no Greek ever be afraid lest it be not right ever to concern ourselves, mortal as we are, with things divine: we must think the very contrary, seeing that God is never either unintelligent or ignorant in any way of our human nature, but he knows that, when he teaches us, we shall follow closely and shall learn what we are taught. And, of course, he knows that he teaches us this very subject, and that we learn both number and numbering. It would, indeed, be the height of unintelligence not to know this; as the saying is, (the divine nature) would truly not know itself if it should feel resentment towards any one who has the ability to learn, and should not rather ungrudgingly rejoice with one who has, by God's help, become good. It would be a long and beautiful story which should tell how, at the time when men first conceived ideas about the gods, the way in which they came to be, the sort of being they became, all and sundry, and the sort of actions they took in hand, they were spoken of in terms repugnant to all wise men, uncivil withal, and not even in the way that the second succession of thinkers used to speak; for with these latter the elements of fire, water, and the rest were spoken of as the chiefest, then later came the miracle of the soul and the superior and worthier motion which was granted to the body so that it could move itself in conditions of heat and cold and such like, without however the soul being supposed able to move the body as well as itself. But now, when we affirm of the soul that, if it be present in the body, there is no wonder that it can move and carry about both the body and itself, our soul does not even suggest to us any doubt on any ground whatever such as that it cannot carry about anything that has weight. Now therefore, too, when we declare that, the soul being the cause of the whole, all good things being what they are, and all bad things being also what they are, but different from the other, it is no wonder that the soul is the cause of all local or other movement, that local or other movement towards the good belongs to the best soul, and movement in the contrary direction to the contrary kind of soul, it necessarily follows that the good has had, and still has, the victory over the bad.

Epinomis, 990 B–C.

The moon describes its own orbit quickest of all, bringing the month and the first full moon; and we must regard as second the sun, which executes its turnings during its own complete circuit, and the planets which keep it company. But, not to repeat over and over again the same remarks about the same things, it is not easy to understand the courses of those other stars which we before described; but we should for this purpose provide persons of such ability as we can find, seeing that much preliminary teaching and practice is required, followed by incessant labour through boyhood and adolescence. It follows that mathematics will be essential.

EUDOXUS (AND CALLIPPUS)

System of concentric spheres

ARISTOTLE, *Metaph.* Λ 8, 1073 b 17–1074 a 15.

EUDOXUS assumed that the sun and moon are moved by three spheres in each case; the first of these is that of the fixed stars, the second moves about the circle which passes through the middle of the signs of the zodiac, while the third moves about a circle latitudinally inclined to the zodiac circle; and, of the oblique circles, that in which the moon moves has a greater latitudinal inclination than that in which the sun moves. The planets are moved by four spheres in each case; the first and second of these are the same as for the sun and moon, the first being the sphere of the fixed stars which carries all the spheres with it, and the second, next in order to it, being the sphere about the circle through the middle of the signs of the zodiac which is common to all the planets; the third is, in all cases, a sphere with its poles on the circle through the middle of the signs; the fourth moves about a circle inclined to the middle circle (the equator) of the third sphere; the poles of the third sphere are different for all the planets except Aphrodite and Hermes, but for these two the poles are the same.

Callippus' additions to the system

Callippus agreed with Eudoxus in the position he assigned to the spheres, that is to say, in their arrangement in respect of distances, and he also assigned the same number of spheres as Eudoxus did to Zeus and Kronos respectively, but he thought it necessary to add two more spheres in each case to the sun and moon respectively, if one wishes to account for the phenomena, and one more to each of the other planets.

Aristotle's modification

But it is necessary, if the phenomena are to be produced by all the spheres acting in combination, to assume in the case of each of the planets other spheres fewer by one [than the spheres assigned to it by Eudoxus and Callippus]; these latter spheres are those which unroll, or react on, the others in such a way as to replace the first sphere of the next lower planet in the same position [as if the spheres assigned to the respective planets above it did not exist], for only in this way is it possible for a combined system to produce the motion of the planets. Now the deferent spheres are, first, eight [for Saturn and Jupiter], then twenty-five more [for the sun, the moon, and the three other planets]; and of these only the last set [of five] which carry the planet placed lowest [the moon] do not require any reacting spheres. Thus the reacting spheres for the first two bodies will be six, and for the next four will be sixteen; and the total number of spheres, including the deferent spheres and those which react on them, will be fifty-five. If, however, we choose not to add to the sun and moon the [additional deferent] spheres we mentioned [i.e. the two added to each by Callippus], the total number of the spheres will be forty-seven. So much for the number of the spheres.

SIMPLICIUS, on *De caelo*, p. 488, 18–24, Heib.

And, as Eudemus related in the second book of his astronomical history, and Sosigenes also who herein drew upon Eudemus, Eudoxus of Cnidos was the first of the Greeks to concern himself with hypotheses of this sort, Plato having, as Sosigenes says, set it as a problem to all earnest students of this subject to find what are the uniform and ordered movements by the assumption of which the phenomena in relation to the movements of the planets can be saved.

Ib., p. 496, 23–497, 5, Heib.

The third sphere, which has its poles on the great circle of the second sphere passing through the middle of the signs of the zodiac, and which turns from south to north and from north to south, will carry round with it the fourth sphere which also has the planet attached to it, and will moreover be the cause of the planet's movement in latitude. But not the third sphere only; for, so far as it was on the third sphere (by itself), the planet would actually have arrived at the poles of the zodiac circle, and would have come near to the poles of the universe; but, as things are, the fourth sphere, which turns about the poles of the inclined circle carrying the planet and rotates in the opposite sense to the third, i.e. from east to west, but in the same period, will prevent any considerable divergence (on the part of the planet) from the zodiac circle, and will cause the planet to describe about this same zodiac circle the curve called by Eudoxus the *hippopede* (horse-fetter), so that the breadth of this curve will be the (maximum) amount of the apparent deviation of the planet in latitude, a view for which Eudoxus has been attacked.

Ib., p. 504, 17–505, 11; 505, 19–506, 3, Heib.

"Nevertheless the theories of Eudoxus and his followers fail to save the phenomena, and not only those which were first noticed at a later date, but even those which were before known and actually accepted by the authors themselves. What need is there for me to mention the generality of these, some of which, after Eudoxus had failed to account for them, Callippus tried to save—if indeed we can regard him as so far successful? I confine myself to one fact which is actually evident to the eye; this fact no one before Autolycus of Pitane even tried to explain by means of hypotheses, and not even Autolycus was able to do so, as clearly appears from his controversy with Aristotherus. I refer to the fact that the planets appear at times to be near to us and at times to have receded. This is indeed obvious to our eyes in the case of some of them; for the star called after Aphrodite and also the star of Ares seem, in the middle of their retrogradations, to be many times as large, so much so that the star of Aphrodite actually makes bodies cast shadows on moonless nights. The moon also, even in the perception of our eye, is clearly not always at the same distance from us, because it does not always seem to be of the same size under the same conditions as to medium. The same fact is, moreover, confirmed if we observe the moon by means of an instrument; for it is at one time a disk of eleven finger-breadths, and again at another time a disk of twelve finger-breadths, which when placed at the same distance from the observer hides the moon (exactly) so that his eye does not see it. In addition to this, there is evidence for the truth of what I have stated in the observed facts with regard to total eclipses of the sun; for when the centre of the sun, the centre of the moon, and our eye happen to be in one straight line, what is seen is not always alike; but at one time the cone which comprehends the moon and has its vertex at our eye comprehends the sun itself at the same time, and the sun even remains invisible to us for a certain time, while again at another time this is so far from being the case that a rim of a certain breadth on the outside edge is left visible all round it at the middle of the duration of the eclipse. Hence we must conclude that the apparent difference in the sizes of the two bodies observed under the same atmospheric conditions is due to the inequality of their distances (at different times). . . . But indeed this inequality in the distances of each star at different times cannot even be said to have been unknown to the authors of the concentric theory themselves. For Polemarchus of Cyzicus appears to be aware of it, but to minimize it as being imperceptible, because he preferred the theory which placed the spheres themselves about the very centre in the universe. Aristotle, too, shows that he is conscious of it when, in the *Physical Problems*, he discusses objections to the hypotheses of astronomers arising from the fact that even the sizes of the planets do not appear to be the same always. In this respect Aristotle was not altogether satisfied with the revolving spheres, although the supposition that, being concentric with the universe, they move about its centre, attracted him." Again, it is clear from what he says in Book Λ of the *Metaphysics* that he thought that the facts about the movements of the planets had not been sufficiently explained by the astronomers who came before him or were contemporary with him. At all events, we find him using language of this sort: "(on the question, how many in number these movements of the planets are), we must, for the present, content ourselves with repeating what some of the mathematicians say, in order that we may form a notion and our mind may have a certain definite number to apprehend; but for the rest we must, in part, make researches for ourselves, and in part inquire of other investigators, and, if those who study these questions reach conclusions different from the views now put forward, we must, while respecting both, give our adherence to those which are the more correct."

ARISTOTLE

Motion and the prime movent

Aristotle, *Physics*, VIII, 10, 267 a 21–b 9, b 17–26.

Since in the world of being there must be a continuous motion, and this is one, and the one motion must belong to some magnitude (for that which has no magnitude does not move) and must be a movement of one magnitude caused by one movent—otherwise it will not be continuous, but will consist of one movement after another in succession and so will be divided—and the movent, if it is one, must, while causing motion, itself be either in motion or motionless—if now the movent is being moved itself, it must follow (the moved) and suffer change, and it must, at the same time, be moved by something. In the end, therefore, the series must stop, and we shall ultimately arrive at a motion set up by an unmoved movent. This last need not change with what it moves, but it will be capable of causing motion without end—setting up motion in this kind is effortless—and this motion is either the only uniform motion or the most uniform there is; for here the movent suffers no change. But neither must the object moved by it suffer change in relation to that movent if the movement is to be uniform. The movent, therefore, must be either at the centre or on the circumference, for these are the first principles. Now that which is nearest to the movent is moved most swiftly. But the motion of the circumference is the swiftest. Therefore the movent must be located *there*. . . .

After these explanations it is manifest that the primary and unmoved movent cannot possibly have any magnitude. For if it has magnitude, it must be either finite or infinite. Now that there cannot be any infinite magnitude has been proved earlier in the *Physics* (III, 5); and that it is impossible that that which is finite can have infinite potency, and impossible that anything can be moved by a finite thing for an infinite time, has now been established. But the prime movent causes eternal movement lasting for an infinite time. It is manifest, therefore, that it is indivisible, without parts, and has no magnitude.

Aristotle, *Metaph.* Λ 8, 1073 a 23–b 17.

The first principle and the first of existing things is not movable, either in itself or contingently, but it sets up the primary eternal motion, which is one. And since that which is moved must be moved by something, and the prime movent must be in itself immovable, and the eternal motion must be caused by something eternal, and one motion by one movent—but, besides the simple revolution of the universe, which we affirm to be caused by the primary immovable substance, we observe other movements, those of the planets, which are eternal (for the body which has circular motion is eternal and unresting: this has been shown in the Physical Treatises)—it follows that each of these motions also must be caused by a substance immovable in itself and eternal. For the constitution of the stars is eternal, being a kind of substance, and the movent is eternal and prior to that which is moved, and that which is prior to a substance must be a substance. It is manifest, therefore, that there must exist, to the same number as that of the movements of the stars, substances which are by nature eternal, immovable in themselves, and, for the reason before explained, having no magnitude. That the movents, then, are substances, and that among them one is first and another second in an order which is the same as that of the movements of the stars (planets), is manifest. When we come to the number of the revolutions, we are at a point where we must draw upon that one of the mathematical sciences which is most akin to philosophy, namely astronomy, for this concerns itself with substance which, though perceptible to sense, is eternal, whereas the others, like the theory of numbers and geometry, deal with no substance. Now that the movements are more in number than the moving bodies is manifest to those who are even moderately versed in the subject; for each of the planets has more than one movement. On the question how many the movements are, we will now quote, for the purpose of giving some notion of the problem, what certain of the mathematicians say, in order that the mind may have some definite number to conceive of; as for the rest, we must in part make researches for ourselves, and in part inquire of other investigators; and if those who study these things reach conclusions different from what we put forward, we should give credit to both, but follow the more correct.

The stars and the heaven : shape, motions, distances, and speeds : supposed "harmony"

De caelo, II, 289 a 11–291 b 23.

It will be proper, in the next place, to speak of what we call stars, their constituents, their shapes, and their movements. It is most reasonable, and most in accord with what has been said, to regard each of the stars as made out of that body in which their spatial movement takes place, since we said that there was a certain element to which circular motion was natural (I, 2, 3). Just in the same way those who make them to be of fire do so because they declare that the body above us is fire, implying that it is reasonable that anything should be constructed out of that in which it has its being. The heat and light which come from the stars arise through friction with the air which they thrust aside by their motion. For it is a natural effect of motion to set things on fire, such as wood, stones, and iron. This would even more reasonably be expected of that which is nearer to fire, and air is nearer to fire. A case in point is that of flying missiles; for these are themselves set on fire to such an extent that those made of lead are melted, and, since they are themselves set on fire, it follows that the air round about them must be affected in the same way. Now these missiles are made hot because they are carried along in air, which owing to the concussion is turned into fire by the motion. But each of the bodies in the heaven is carried round in the sphere, in such a way that, although the bodies themselves are not set on fire, the air which is under the sphere formed by the circularly revolving body must, as that sphere revolves, be made hot, and this happens most of all in the place where the sun is attached to it. Hence, when the sun approaches nearer, and when it rises high and is above our heads, its heat is felt. So much by way of proof that the stars are not made of fire, and do not move in a medium of fire.

Seeing that the stars on the one hand, and the whole heaven on the other, appear to change their positions, there are three possible hypotheses, namely (1) that both are at rest, (2) that both are in motion, (3) that one is at rest and the other in motion. Now it is impossible that both should be at rest, granted that the earth is at rest; for on this hypothesis the observed phenomena could not have taken place. Let us assume that the earth is at rest. Two alternatives then remain: either the stars and the heaven both move or one moves and the other is at rest. If now both move, it is absurd to suppose that the stars and the circles have the same speeds; yet each of the stars *must* have the same speed as the circle in which it is carried, since the stars appear to return again to the same position at the same time as the circles do, and it follows that the star has traversed its own circle, and the circle has in completing its own revolution traversed its own circumference, in the same time. But it is not reasonable that the speeds of the stars and the sizes of the circles should have the same ratio. For, while it is in no way absurd, nay, it is essential that the speeds of circles should be proportional to their sizes, it is the reverse of reasonable to expect this of the several stars in them. Suppose, on the one hand, that that which is carried in the larger circle necessarily moves more swiftly; then it is clear that, if the stars be transferred to each other's circles, the one will be swifter, and the other slower, than before, and this would mean that they have not a movement of their own but are carried by the circles. On the other hand, if their movements had spontaneously coincided, even then it would not be reasonable to expect that greater size in the circles and greater swiftness in the revolution of the star upon it would in all cases go together; that it should be so with one or two would not be impossible, but to assume it for all stars alike is fantastic. Besides, there is in things natural no happening by chance, nor is that which applies to everything everywhere the result of chance.

But once more, if the circles remain stationary, while the stars themselves move, this assumption also is equally absurd, for the result will be that the outer stars move more swiftly and the speeds correspond to the sizes of the circles.

Since, then, it is not reasonable to suppose either that both move or that the star alone moves, there remains only the hypothesis that the circles move, and that the stars are [themselves] at rest, but move through being fastened to the circles; only on this supposition does no absurdity arise. For it is reasonable that, when circles are attached about the same centre, the speed of the greater should be swifter (as in other cases the greater body describes its proper course the more swiftly, so it is with circular motions, for, of the segments cut off by the radii, the segment of the greater circle is greater, so that the greater circle may reasonably be supposed to be carried round in the same time as the lesser), and it is this reason, combined with the fact already proved, that the whole universe is continuous, which accounts for the heaven not breaking asunder.

Again, since the stars are spherical in shape, as others maintain, and as we can agree to hold, seeing that we generate them from the body which we described (the "aether"), and since that which is spherical has two kinds of movement natural to it, namely rolling and spinning, it follows that, if the stars had a movement of their own, it would be one of these two. In fact, however, they seem to have neither. For, if they had whirled or spun, they would have remained in the same place, and would not have changed their position, as they appear to do and as every one maintains. Again, it would be reasonable that all of them should have the same motion; but the only one of the heavenly bodies that appears to rotate is the sun at rising or setting, and the cause of this appearance is not the sun itself, but the fact that it is so far removed from our eyes; for our sight, when at long range, wavers on account of its weakness. This is perhaps the reason why the stars, which are fastened (to the sphere), appear to twinkle, while the planets do not twinkle; for the planets are near, so that the visual ray when it reaches them is within its powers, whereas, when directed to the fixed stars, it quivers on account of the distance, being strained too far. Its quivering makes the movement seem to belong to the star; for it makes no difference whether what is set in motion is the visual ray or the object seen. Further, it is manifest that neither do the stars *roll*; for that which rolls must rotate, whereas the so-called face in the moon is always visible to us. Accordingly, since, if they move of themselves, it is reasonable that they should move with their proper motions, but nevertheless they do not appear to be so moved, it is manifest that they cannot move of themselves. Besides, on that hypothesis, it is absurd that nature has given them no organ to serve for motion. Nature does nothing at haphazard, nor can she be supposed to look after living beings but to overlook objects so precious as the stars; yet in their case she seems of set purpose, as it were, to have taken away every means whereby they might have propelled themselves, and to have made them as far removed as possible from the creatures which have organs for motion. Hence we may reasonably suppose that the whole heaven is spherical in shape, and that each of the stars is so too. For, of all figures, the sphere is the most adapted for motion in the same place (since it admits of the highest speed in its motion and best secures retention of exactly the same position), but is the least adapted for motion forwards, since it is least like the things which can move of themselves. For it has nothing tacked on or projecting as a rectilineal figure has, but is the furthest removed in shape from the bodies which are capable of travelling.

Seeing then that the motion of the heaven must be motion in one and the same place, and that the stars do not have to move forward of themselves, both the heaven and the stars may reasonably be supposed to be spherical, since that supposition best accounts for the one moving and the other being at rest.

It is clear from this that, when it is asserted that the movement of the stars produces a harmony, the sounds which they make being in accord, the statement, although it is a brilliant and remarkable suggestion on the part of its authors, does not represent the truth. I refer to the view of those who think it inevitable that, when bodies of such size move, they must produce a sound; this, they argue, is observed even of bodies within our experience, which neither possess equal mass nor move with the same speed; hence, when the sun and moon, and the stars, which are so many and of such size, move with such a velocity, it is impossible that they should not produce a sound of intolerable loudness. Supposing, then, that this is the case, and that the velocities depending on their distances correspond to the ratios representing chords, they say that the tones produced by the stars moving in a circle are in harmony. But, as it must seem absurd that we should not all hear these tones, they say that the reason of this is that the sound is already going on at the moment we are born, so that it is not distinguishable by contrast with its opposite, silence; for the distinction between vocal sound and silence involves comparison between them; thus a coppersmith is apparently indifferent to noise through being accustomed to it, and so it must be with men in general.

Now this doctrine is, as we said before, elegant and

poetical, but it is impossible that the case should be so. It is impossible, not only that we should hear nothing, a fact for which attempts are made to account, but also that no effect should be produced on us even apart from sensation. For excessively powerful noises shiver the masses even of inanimate bodies, as the noise of thunder cracks stones and the most rigid of bodies. But if bodies of such size were moving, and if the sound transmitted were proportionate to the size of the moving body, it would be an amount of noise many times multiplied that would reach us, and the degree of its violence would be overpowering. But it is, in fact, quite reasonable that we should hear no sound, and that bodies should not appear to suffer any violent effects, because there *is* no sound. The reason of this is plain, and confirms the truth of my remarks; nay, the very difficulty which made the Pythagoreans argue that the moving bodies caused a harmony is evidence in my favour. For things which move of themselves cause sound and concussion, but things which are fastened to or inherent in a thing which moves, such as the component parts in a ship, cannot make a sound any more than the ship itself does when moving in a river. Yet you could, in this case, use the selfsame arguments, and say that it was absurd that the mast and the stern of a vessel of such size should not make a noise, or for that matter the ship itself. A thing moving in something which is not moved makes a sound; but a thing moving in something which moves continuously and causes no concussion, cannot possibly make a sound. Accordingly we must here maintain that, if the bodies of the stars had moved in a mass of air or of fire diffused through the whole universe, as every one holds, they must necessarily have produced a sound of excessive loudness, and in that event it must have reached us and shivered things. As however this does not appear to happen, none of the stars can have either the motion of an animate being or a motion forcibly imparted to it. It is as if nature had foreseen what would follow, namely that, if the motion had not been what it actually is, nothing in the place where we live could have remained in the same condition.

We have now proved that the stars are spherical in shape, and do not move of their own motion.

As regards the order of the heavenly bodies, their position in respect of some being further forward and others further back, and their relation to one another in respect of their distances, we should consult the works of astronomers, where the subject has been adequately dealt with. It is found that their several movements, in so far as some are quicker and others slower, correspond to their distances. For, since it is our hypothesis that the outermost revolution of the heaven is simple and is the swiftest, while those of the other heavenly bodies are slower and have more than one component, for each of them moves, in a sense contrary to the motion of the heaven, in a circle of its own, it is reasonable that that which is nearest to the simple and primary revolution should describe its own circle in the longest time, and that which is furthest from it in the shortest time, and, of the rest, that which is continually nearer should describe its circle in a longer, and that which is further away in a shorter, time. For the revolution nearest to the outermost is dominated the most, and that which is furthest away the least, owing to the difference in distance; while those which are intermediate vary according to the distances, as the mathematicians have proved.

That the shape of each star is spherical is the most reasonable view to take. For, since it has been shown that it is not in their nature to move of themselves, and Nature does nothing without reason or uselessly, it is clear that she gave to the immovable bodies the shape least adapted for movement. The least adapted is the sphere, because it has no organs to move with; hence the mass of each star must clearly be spherical. Again, what applies to one applies equally to all. The moon in particular is shown by ocular evidence to be spherical; were this not so, she would not, in waxing and waning, have taken the form mostly of a crescent or of a doubly-convex figure, and only once have appeared halved. It is also proved by astronomical considerations; for, were it not so, the eclipses of the sun would not have shown a crescent-shaped figure. If, then, one heavenly body is spherical in shape, it is clear that the rest will also be spherical.

De caelo, II, 291 b 29–292 a 9.

What can be the reason why it is not always the bodies which are at a greater distance from the primary motion [the daily rotation of the sphere of the fixed stars] that are moved by more movements, but it is the middle bodies which have most movements? For it would appear reasonable that, as the primary body [the sphere of the fixed stars] has one motion only, the nearest body to it should be moved by the next fewest movements, say two, the next to that by three, or in accordance with some other arrangement of this sort.

But in practice the opposite is the case; for the sun and moon are moved by fewer movements than some of the planets, and yet the latter are further away from the centre and nearer the primary body [the sphere of the fixed stars] than the sun and moon are. In the case of some planets this is even observable by the eye; for, at a time when the moon was halved, we have seen the star of Ares go behind it and become hidden by the dark portion of the moon, and then come out at the bright side of it. And the Egyptians and Babylonians of old, whose observations go back a great many years, and from whom we have a number of accepted facts relating to each of the stars, tell us of similar occultations of the other stars.

The earth: its position, shape, rest or motion: historical sketch

De caelo, II, 11, 293 b 30–294 b 30 (*continuation of passage on pp. 30–31*).

Some say that the earth, while placed at the centre, "rolls," that is, moves, about the axis stretching through the whole universe, as we read in the *Timaeus*. Similarly, there is also dispute about its shape. Some think it spherical, others flat and of the shape of a tambourine. The latter adduce as evidence the fact that, when the sun is setting and rising, the line demarcating the portion hidden by the earth appears to be straight and not circular, the assumption being that, if the earth had been spherical, the line of section must be circular. In this, however, they fail to take account of the distance of the sun from the earth and the size of the earth's circumference, an arc of which, projected on to circles appearing to be small, appears at such an immense distance to be a straight line. This appearance ought not to make these thinkers feel any doubt that the mass of the earth is spherical. But they use a further argument and say that, because the earth is at rest, it must have this [flat] shape. And, of course, there are many varieties of view regarding the question of the earth's motion or rest.

Difficulties must necessarily occur to every one; for it would be a mark of a very untroubling intellect not to feel surprise that a small fragment of the earth, if it be lifted up and then dropped, moves (i.e. falls) and will not stay where it is put, and the greater the fragment is, the swifter the motion, and yet that the earth as a whole, if lifted up and let go, should not move at all. As it is, however, though its weight is so enormous, it remains at rest. Further, if from under the aforesaid portions of the earth when already in motion, but before they fall, the earth were taken away, they would continue to be carried downwards, there being nothing to stop them. The difficulty has naturally become a stock problem in philosophy for every one. But one may well wonder that the solutions of it should not be thought even more absurd than the question. Some have been induced by the above difficulty to assert that the under-portion of the earth is infinite, saying, as Xenophanes of Colophon does, that "its roots extend to infinity," to save themselves the trouble of investigating the cause of its being at rest. Hence Empedocles' rebuke in the words, "If the depths of the earth are limitless, and limitless the vast aether above it, as has been said by the tongues of many, and vainly spouted forth from the mouths of men who have seen little of the whole."

Others say that the earth lies on water. This, the most ancient theory that has come down to us, is that attributed to Thales of Miletus, the idea being that it remains where it is because it floats like a log or any similar object, for none of these things are so constituted as to rest upon air, though they do so on water. As though the same argument as applies to the earth would not apply to the water carrying the earth! For neither is water so constituted as to remain in mid-air, but it requires something to rest upon. Again, as air is lighter than water, so is water lighter than earth; how then can they suppose that the lighter remains below what is naturally heavier? Further, if the earth as a whole is so constituted as to rest upon water, it is clear that so must any portion of it: this, however, does not appear to be the case, but any casual fragment of it sinks to the bottom, and the larger the fragment the quicker. The fact is that these thinkers have got some way in investigating the problem, but not so far as it is possible to probe it. We are all prone, in conducting our researches, to look, not to the subject itself, but to what our opponents say; a man inquiring into anything by himself goes just so far as the point where he no longer has anything to oppose to his own argument. Hence one who proposes to make an investigation of any value must be capable of stating the objections proper to the genus, and he can only be this when he has considered all the differences.

Anaximenes, Anaxagoras, and Democritus say that its flatness is what makes the earth remain at rest; for it does not cut the air below it, but acts like a lid to it, and this appears to be characteristic of those bodies which possess breadth. Such bodies are, as we know, not easily displaced by winds, because of the resistance they offer. The philosophers in question assert that the earth, by its breadth, offers the same resistance to the air below it, and that the air, on the other hand, not having sufficient space to move from its position, because it is below the earth, remains in one mass, just as the water does in water-clocks. And they produce much evidence to show that air, when cut off and remaining at rest, is capable of bearing a great weight.

Now, first of all, if the shape of the earth is not flat, the reason why it remains at rest cannot be its flatness. And indeed their statements imply that the cause of its remaining stationary is not its breadth, but rather its size. The reason of the air being in a confined space, and so unable to find a passage, is its great bulk; and it has this great bulk because the body by which it is shut off, namely the earth, is of great size. This will, therefore, still be the case even if the earth is spherical, provided it is of the required size; it will, according to their argument, equally remain at rest. . . .

Ib., 295 a 13–296 b 26.

All who generate the heavens agree in saying that the earth came together in the centre [as the result of a "whirling" motion, such as that of eddies in water or in the air]; but they have to find the reason why it remains there. Some argue, in the way we mentioned, that its breadth and its size are the reason. Others, like Empedocles, say that the revolution of the heaven moving circularly round about the earth at a higher speed prevents the earth from moving, just like the water in a cup, which may be swung round in a circle so fast that the water is often actually below the bronze of the cup, and yet, for the same reason, is not carried down with the motion natural to it (i.e. is not spilt). But, suppose that neither the "whirling" nor its flatness prevents the earth from moving, and that the air is withdrawn, where in the world is it to move to? Compulsion brought it to the centre, and it is under compulsion that it remains there. But there must be some motion which is natural to it. Is this motion up or down, or where? It must have some motion; and if it is no more up than down, and the air above does not prevent its moving upwards, neither will the air below it prevent its moving downwards. For the same causes must of necessity produce the same effects on the same things.

Further, we might use another argument against Empedocles. When the elements were separated off by Strife, why did the earth remain still? He will surely not argue that the "whirling" was *then* the cause. It is absurd, too, not to realize that, while the parts of the earth originally came together in the centre because of the "whirling," we have to explain why all heavy things *now* move towards it, for surely the whirling does not come near us? Further, why does fire move upwards? Surely this is not because of the "whirling"? And if fire has a natural motion in some direction, we must clearly suppose that the earth also has. Again, it is surely not by the "whirling" that the heavy and light have been distinguished; but it was pre-existing heavy and light things of which the motion caused some to go to the centre and others to float on the surface. Heavy and light, therefore, existed before the whirling began. By what, then, were they distinguished, and how was it their nature to move and where? For in anything infinite there cannot be any up and down, but it is by these that the heavy and light are distinguished.

Most philosophers concern themselves with causes of this kind. But there are others, such as Anaximander among the ancients, who allege indifference as the cause of the earth's remaining at rest. The argument is that it is no more appropriate to that which is established at the centre and is similarly situated with regard to the extremities to move upwards than downwards or sideways; and a thing cannot move in contrary directions at the same time; it must, therefore, necessarily remain stationary. This argument is clever, but it is not true; for, on this showing, anything whatever which is put at the centre must necessarily remain there, so that even fire will rest at the centre, for the argument does not depend on anything peculiar to the earth. But this does not follow. For apparently not only does the earth remain at the centre, but it moves towards the centre. To whatever place any fragment of the earth moves, to that place the whole earth must also necessarily move; and to whatever place it naturally moves, there it will naturally rest. The reason, then, is not that it is similarly situated with regard to the extremities; this is common to everything, but movement to the centre is peculiar to earth. Again, it is absurd to inquire what can be the reason why the earth remains stationary at the centre, and not also to inquire why fire remains at the extremity. If the extreme place is natural to fire, clearly there must be some place natural to earth. But, even if its present place (the centre) is not natural to the earth, and the reason of its remaining there is the necessity of the principle of indifference — this, by the way, is analogous to the argument that a hair, however strong the tension applied to it, provided it is exactly the same throughout its length, will never break, or the argument about the man who is hungry and thirsty to a degree, but both equally, and is placed at equal distances from food and drink, and who, it is alleged, would therefore have to remain where he is—even so, these thinkers have still to find the reason why fire rests at the extremities.

It is strange, too, to inquire about rest, but not to inquire about the motion of bodies. I mean the question why, if there is nothing to stop it, one thing moves upwards and another to the centre. However, the argument about indifference is not true. Incidentally it is true that anything to which motion in one direction is not more appropriate than motion in another must necessarily remain at the centre. But, so far as their argument goes, it will not remain at rest, but will move, in fragments though not as a whole. For the same argument will apply to fire also; it must necessarily (on the hypothesis of indifference) remain where it is put just like the earth, since it will be similarly related to any one whatever of the extreme points. But nevertheless it will move, as we see it doing if nothing prevents it, away from the centre to the extremity, though it will not move as a whole to one point (this fact [namely that it does not move as a whole towards one point] is all that necessarily follows from the argument about indifference), but a proportionate amount to each proportionate part of the extremity, say (for instance) one fourth part to one fourth part of the enveloping (sphere); and no body [or part of a body] is a point. And, just as anything may be condensed and come together from a larger space into a smaller, so it may become rarer, and take up a larger space instead of a smaller; hence, for anything that the argument from indifference can say to the contrary, the earth would have executed this sort of movement away from the centre, if the centre had not been its natural place. The views, then, that have been held about the shape of the earth, and about its place and the question of its rest or motion, are roughly such as we have described.

Let us now, speaking for ourselves, first consider whether the earth is in motion or at rest, for, as we said, some make it one of the stars, while others, though they place it at the centre, say that it "rolls," that is, moves, about the axis in the midst. That this is impossible is clear if we assume as a principle that, if the earth moves, whether it be away from the centre or at the centre, it must necessarily make this movement under compulsion. It cannot belong to the earth itself; for, if it did, each of its parts would also have had this motion; but, as it is, they all move in a straight line towards the centre. Hence the motion of the earth, being due to compulsion, and not natural to it, cannot be eternal, whereas, of course, the ordering of the universe is eternal. Further, all things which move in a circle, except the first (outermost) sphere, appear to be left behind and to have more than one movement; hence the earth, too, whether it moves about the centre or in its position at the centre, must have two movements. Now if this occurred, it would follow that the stars would exhibit passings and turnings-back. This does not, however, appear to be the case, but the same stars always rise and set at the same places on the earth.

Again, the natural movement, both of parts of the earth and of the earth as a whole, is towards the centre of the universe—this is why the earth now actually lies at the centre—but it might be questioned, seeing that both have the same centre, to which centre it is that things which have weight, and the parts of the earth, naturally move; in other words, whether they move to the one centre because it is the centre of the universe, or because it is the centre of the earth. It must be the centre of the universe to which they move, for light things and fire, which move the contrary way to things which have weight, move towards the extremities of the space which surrounds the said centre. It happens that the same point is the centre both of the earth and of the universe. The motion, then, is towards the centre of the earth, but only accidentally, by virtue of the earth's centre being the centre of the universe. That heavy bodies do move towards the centre of the earth is indicated by the fact that weights moving towards the earth do not move in parallel directions, but at similar (i.e. equal) angles (i.e. at right angles to the spherical surface of the earth), so that they move to one centre, which is the centre of the earth. It is manifest, therefore, that the earth is at the centre and immovable, both for the reasons aforesaid, and also because weights forcibly thrown plumb upwards move back again to the same place, even if the force exerted be sufficient to throw them an infinite distance. It is clear from these considerations that the earth neither moves nor lies away from the centre.

Ib., 297 a 2–b 14.

We have evidence for our view in what the mathematicians say about astronomy. For the phenomena observed as changes take place in the figures by which the arrangement of the stars is marked out occur as they would on the assumption that the earth is situated at the centre [The answer to this is, of course, How do you know?]. So much for the position of the earth and the question in what state it is in regard to rest or motion.

Spherical shape of the earth

Its shape must, of necessity, be spherical. For every one of its parts has weight right down to the centre. The jostling of the lesser part by the greater cannot result in a wave-like formation, rather it suffers greater compression, and part must come together with part until the centre is reached. The process must be understood to be the same as would have taken place if the earth had come into being in the way described by some of the natural philosophers. Only they allege compulsion as the reason for the downward movement, and we had better give the true reason, and say that that which has weight has a natural tendency to move towards the centre. When the mixture was in the potential stage, the things that were being separated off were moving uniformly from all directions towards the centre. Now, whether the parts which came together from all directions to the centre were evenly distributed at the extremities, or were otherwise disposed, the result would be the same. Given that the parts came together in similar quantities from all parts of the extremities to the one centre, it is manifest that the resulting mass must be similar on all sides; for, if an equal amount be added all over, the extremities must necessarily be equidistant from the centre, and it is this form which belongs to a sphere. But neither will it affect the argument if the coming together of the parts was not uniform from all directions. For the greater must continually push forward the lesser which is in front of it, both having the tendency to move as far as the centre, and the heavier, therefore, pushing forward the lesser weight until the centre is reached. Another difficulty that may be felt can be solved in the same way. Suppose that, the earth being at the centre and spherical in shape, a weight many times as great were added to one hemisphere; the centre of the whole would then not be the same as the centre of the earth; accordingly, either it will not remain at the centre, or, if it does, it will have to be at rest without having as its centre the place to which it is still its natural tendency to move. The difficulty is as stated; but it is easy to see the answer if we make a little effort, and first define in what sense we assume that a magnitude which has weight, whatever its size, moves to the centre. Clearly it will not move just until its extremity coincides with the centre (and no longer), but the greater part of it must prevail until the body's own centre occupies the centre, for its tendency persists until this happens. It makes no difference whether we assert this of a clod or any portion of the earth, or of the earth as a whole; what we have stated does not depend on any question of smallness or greatness, but applies to all bodies whatever that have a natural tendency towards the centre, so that, whether the earth began to move as a whole from somewhere, or bit by bit, it must necessarily move until its occupation of the centre is similar in all directions, the lesser portions being forced by the greater into equality (i.e. equal distribution) by virtue of the forward thrust due to the tendency.

HERACLIDES OF PONTUS

Rotation of the earth on its axis

SIMPLICIUS, on Arist. *De caelo*, II, 13, 293 b 30; p. 519, 9–11, Heib.

BUT Heraclides of Pontus, by supposing that the earth is in the centre, and rotates, while the heaven is at rest, thought in this way to save the phenomena.

Ib., on *De caelo*, II, 7, 289 b 1; p. 444, 31–445, 5, Heib.

He (Aristotle) thought it right to take account of the hypothesis that *both* (i.e. the stars and the whole heaven) are at rest—although it would appear impossible to account for their apparent change of position on the assumption that both are at rest—because there have been some, like Heraclides of Pontus and Aristarchus, who supposed that the phenomena can be saved if the heaven and the stars are at rest, while the earth moves about the poles of the equinoctial circle from the west (to the east), completing one revolution each day, approximately; the "approximately" is added because of the daily motion of the sun to the extent of one degree. For, of course, if the earth did not move at all, as Aristotle will later show to be the case, although he here assumes, for the sake of argument, that it does, it would be impossible for the phenomena to be saved on the supposition that the heaven and the stars are at rest.

Aëtius, III, 13, 3.

Heraclides of Pontus and Ecphantus the Pythagorean move the earth, not however in the sense of translation, but in the sense of rotation, like a wheel fixed on an axis, from west to east, about its own centre.

Motion of Mercury and Venus round the sun

Chalcidius, on the *Timaeus*, c. 110.

Lastly, Heraclides Ponticus, when describing the circle of Lucifer, as well as that of the sun, and giving the two circles one centre and one middle, showed how Lucifer is sometimes above, sometimes below the sun. For he says that the position of the sun, the moon, Lucifer, and all the planets, wherever they are, is defined by one line passing from the centre of the earth to that of the particular heavenly body. There will then be one straight line drawn from the centre of the earth showing the position of the sun, and there will equally be two other straight lines to the right and left of it respectively and distant 50° from it, and 100° from each other, the line nearest to the east showing the position of Lucifer or the Morning Star when it is furthest from the sun and near the eastern regions, a position in virtue of which it then receives the name of the Evening Star, because it appears in the east at evening after the setting of the sun.

VITRUVIUS, *De architectura*, IX, 1 (4), 6.

The stars of Mercury and Venus make their retrograde motions and retardations about the rays of the sun, forming by their courses a wreath or crown about the sun itself as centre. It is also owing to this circling that they linger at their stationary points in the spaces occupied by the signs. (Here and in the following passage the name of Heraclides is not given.)

MARTIANUS CAPELLA, *De nuptiis Philologiae et Mercurii*, VIII, 880, 882.

For, although Venus and Mercury are seen to rise and set daily, their orbits do not encircle the earth at all, but circle round the sun in a freer motion. In fact, they make the sun the centre of their circles, so that they are sometimes carried above it, at other times below it and nearer to the earth, and Venus diverges from the sun by the breadth of one sign and a half [45°]. But, when they are above the sun, Mercury is the nearer to the earth, and when they are below the sun, Venus is the nearer, as it circles in a greater and wider-spread orbit. . . .

The circles of Mercury and Venus I have above described as epicycles. That is, they do not include the round earth within their own orbit, but revolve laterally to it in a certain way.

EUCLID

Phaenomena, Preface.

SINCE the fixed stars are always seen to rise from the same place and to set at the same place, and those which rise at the same time are seen always to rise at the same time, and those which set at the same time always to set at the same time, and these stars in their courses from rising to setting remain always at the same distances from one another, while this can only happen with objects moving with circular motion, when the eye (of the observer) is equally distant in all directions from the circumference, as is proved in the Optics, we must assume that the (fixed) stars move circularly, and are fastened in one body, while the eye is equidistant from the circumferences of the circles. But a certain star is seen between the Bears which does not change from place to place, but turns about in the position where it is. And, since this star appears to be equidistant in all directions from the circumferences of the circles in which the rest of the stars move, we must assume that the circles are all parallel, so that all the fixed stars move in parallel circles having for one pole the aforesaid star.

Now some of the stars are seen neither to rise nor to set because they are borne on circles which are high up, and are called "always visible" circles. These stars are those which come next to the visible pole and reach as far as the arctic circle. And, of these stars, those nearer the pole move on smaller circles, and those on the arctic circle on the greatest circle, the latter stars appearing actually to graze the horizon.

But all the stars which follow on these towards the south are all seen to rise and to set because their circles are not wholly above the earth, but part of them is above, and the remainder below, the earth. And of the segments of the several circles that are above the earth, that appears larger which is nearer to the greatest of the always-visible circles, while of the segments under the earth that which is nearer to the said circle is less, because the time taken by the motion under the earth of the stars which are on the said circle is the least, and the time taken by their motion above the earth is the greatest, while, for the stars on the circles which are continually further from the said circle, the time taken by their motion above the earth is continually less, and the time taken by their motion under the earth greater; the motion above the earth takes the least time, and the motion under the earth the greatest time, in the case of the stars which are nearest the south. The stars on the circle which is the middle one of all the circles, appear to take equal times to complete their motion above the earth and their motion under the earth respectively, and hence we call this circle the "equinoctial"; and those stars which are on circles equidistant from the equinoctial circle take equal times to describe the alternate segments; thus the segments above the earth in the northerly direction are equal to those under the earth in the southerly direction, and the segments above the earth in the southerly direction are equal to those under the earth in the northerly direction; but the sum of the times taken by the motion above the earth and by the motion under the earth continuous with it added together appears to be the same for each circle.

Further, the circle of the Milky Way and the zodiac circle, which are both obliquely inclined to the parallel circles and cut one another, appear in their revolution always to show semicircles above the earth.

On all the aforesaid grounds let us make it our hypothesis that the universe is spherical in shape; for if it had been in the form either of a cylinder or of a cone, the stars taken on the oblique circles bisecting the equinoctial circle would, in their revolution, have seemed to describe, not always equal semicircles, but sometimes a segment greater than a semicircle, and sometimes a segment less than a semicircle. For, *if a cone or a cylinder be cut by a plane not parallel to the base, the section arising is a section of an acute-angled cone, which is like a shield* (an ellipse). Now it is clear that, if such a figure be cut through its centre lengthwise and breadthwise respectively, the segments respectively arising are dissimilar; it is also clear that, even if it be cut in oblique sections through the centre, the segments formed are dissimilar in that case also; but this does not appear to happen in the case of the universe. For all these reasons, the universe must be spherical in shape, and revolve uniformly about its axis, one of the poles of which is above the earth and visible, while the other is under the earth and invisible.

Let the name "horizon" be given to the plane passing through our eye which is produced to the (extremities of the) universe, and separates off the segment which we see above the earth. The horizon is a circle; for, *if a sphere be cut by a plane, the section is a circle.*

Let the name "meridian circle" be given to the circle through the poles which is at right angles to the horizon, and the name "tropics" to the circles which are touched by the circle through the middle of the signs (the zodiac) and which have the same poles as the sphere.

The zodiac circle and the equinoctial circle are great circles; for they bisect one another. For the beginning of the Ram (Aries) and the beginning of the "Claws" (Libra) are diametrically opposite to one another and, both being on the equinoctial circle, the rising of the one and the setting of the other take place in conjunction, since they have between them six of the twelve signs of the zodiac, and two semicircles of the equinoctial circle, respectively, and since both beginnings, being on the equinoctial circle, take the same time to describe, the one its course above the earth, the other its course under the earth. But if a sphere rotates about its own axis, all the points on the surface of the sphere describe, in the same time, similar arcs of the parallel circles on which they are carried; therefore the points in question traverse similar arcs of the equinoctial circle, on one side the arc above the earth, on the other the arc under the earth; therefore the arcs are equal; therefore both are semicircles, for the distance from rising to rising, or from setting to setting, is the whole circle; therefore the zodiac circle and the equinoctial circle bisect one another. But, if in a sphere two circles bisect one another, both of the intersecting circles are great circles; therefore the zodiac circle and the equinoctial circle are great circles.

The horizon, too, is one of the great circles. For it always bisects both the zodiac circle and the equinoctial circle; for it has always six of the twelve signs above the earth, and always a semicircle of the equinoctial circle above the earth; for the stars on the latter circle which rise and set respectively at the same time pass, in the same time, the one its course from rising to setting, the other its course from setting to rising. It is therefore manifest, from what was before proved, that there is always a semicircle of the equinoctial circle above the horizon. But if, in a sphere, a circle remaining fixed bisects any of the great circles which is moving continually, the circle which cuts it is also a great circle; therefore, the horizon is one of the great circles.

The time of a revolution of the universe is the time in which each of the fixed stars passes from one rising to its next rising, or from any place whatever to the same place again. . . .

Proposition 1

The earth is in the middle of the universe, and occupies the place of centre in relation to the universe.

ARISTARCHUS OF SAMOS

ARISTARCHUS, *On the Sizes and Distances of the Sun and Moon.*

(Hypotheses.)

1. *That the moon receives its light from the sun.*
2. *That the earth is in the relation of a point and centre to the sphere in which the moon moves.*
3. *That, when the moon appears to us halved, the great circle which divides the dark and the bright portions of the moon is in the direction of our eye.*
4. *That, when the moon appears to us halved, its distance from the sun is then less than a quadrant by one-thirtieth part of a quadrant.*[1]

[1] i.e. 90° – 3° or 87°. The true value is 89° 50′.

5. *That the breadth of the earth's shadow is that of two moons.*

6. *That the moon subtends one-fifteenth part of a sign of the zodiac.*[1]

(Given these hypotheses) it is proved that:

1. *The distance of the sun from the earth is greater than eighteen times, but less than twenty times, the distance of the moon from the earth*: this follows from the hypothesis about the halved moon.

2. *The diameter of the sun has the same ratio as aforesaid to the diameter of the moon.*

3. *The diameter of the sun has to the diameter of the earth a ratio greater than that which* 19 *has to* 3, *but less than that which* 43 *has to* 6: this follows from the ratio thus discovered between the distances, the hypothesis about the shadow, and the hypothesis that the moon subtends one-fifteenth part of a sign of the zodiac.

Pappus, *Synagoge*, VI, pp. 554, 6–560, Hultsch.

In his book on sizes and distances Aristarchus lays down these six hypotheses:

1. That the moon receives light from the sun.

2. That the earth is in the relation of a point and centre to the sphere in which the moon moves.

3. That, when the moon appears to us halved, the

[1] According to Archimedes, Aristarchus "discovered that the sun appeared to be about 1/720th part of the circle of the zodiac": that is, Aristarchus discovered (evidently at a date later than that of the treatise) the much more correct value of $\frac{1}{2}°$ for the angular diameter of the sun or moon (for he maintained that both had the same angular diameter).

great circle which divides the dark and the bright portions of the moon is in the direction of our eye.

4. That, when the moon appears to us halved, its distance from the sun is then less than a quadrant by one-thirtieth of a quadrant.

5. That the breadth of the (earth's) shadow is that of two moons.

6. That the moon subtends one-fifteenth part of a sign of the zodiac.

Now the first, third, and fourth of these hypotheses practically agree with the assumptions of Hipparchus and Ptolemy. For the moon is illuminated by the sun at all times except during an eclipse, when it becomes devoid of light through passing into the shadow which results from the interception of the sun's light by the earth and which is conical in form; next the (circle) dividing the milk-white portion which owes its colour to the sun shining upon it and the portion which has the ashen colour natural to the moon itself is indistinguishable from a great circle (in the moon) when its positions in relation to the sun cause it to appear halved, at which times (a distance of) very nearly a quadrant on the circle of the zodiac is observed (to separate them); and the said dividing circle is in the direction of our eye, for this plane of the circle if produced will, in fact, pass through our eye in whatever position the moon is when for the first or second time it appears halved.

But, as regards the remaining hypotheses, the aforesaid mathematicians have taken a different view. For, according to them, the earth has the relation of a point and centre, not to the sphere in which the moon moves, but to the sphere of the fixed stars, the breadth of the earth's shadow is not (that) of two moons, nor does the moon's diameter subtend one-fifteenth part of a sign of the zodiac, that is, 2°. According to Hipparchus, on the one hand, the circle described by the moon is measured 650 times by the diameter of the moon, while the earth's shadow is measured by it $2\frac{1}{2}$ times at its mean distance in the conjunctions; in Ptolemy's view, on the other hand, the moon's diameter subtends, when the moon is at its greatest distance, an arc of 0° 31′ 20″, and, when at its least distance, of 0° 35′ 20″, while the diameter [Pappus should have said "radius"] of the circular section of the shadow is, when the moon is at its greatest distance, 0° 40′ 40″, and when the moon is at its least distance, 0° 46′.

Hence it is that the authorities named have come to different conclusions as regards the ratios both of the distances and of the sizes of the sun and moon.

Now Aristarchus, after stating the aforesaid hypotheses, proceeds in a passage which I will quote, word for word:

"We are now in a position to prove that the distance of the sun from the earth is greater than 18 times, but less than 20 times, the distance of the moon, and the diameter of the sun also has the same ratio to the diameter of the moon: this follows from the hypothesis about the halved (moon). Again, we can prove that the diameter of the sun is to the diameter of the earth in a greater ratio than that which 19 has to 3, but in a less ratio than that which 43 has to 6; this follows from the ratio thus discovered as regards the distances, from the hypothesis about the shadow, and from the hypothesis that the moon subtends one-fifteenth part of a sign of the zodiac."

He says: "We are in a position to prove that the

distances," etc., implying that he will prove the properties after giving such preliminary lemmas as are of use for proving them. As the result of the whole investigation he concludes that:

(1) the sun has to the earth a greater ratio than that which 6859 has to 27, but a less ratio than that which 79507 has to 216;

(2) the diameter of the earth is to the diameter of the moon in a greater ratio than that which 108 has to 43, but in a less ratio than that which 60 has to 19; and

(3) the earth is to the moon in a greater ratio than that which 1259712 has to 79507, but in a less ratio than that which 216000 has to 6859.

But Ptolemy proved in the fifth book of his Syntaxis that, if the radius of the earth is taken as the unit, the greatest distance of the moon at the conjunctions is $64\frac{10}{60}$ of such units, the greatest distance of the sun 1210, the radius of the moon $\frac{17}{60}+\frac{33}{60^2}$, the radius of the sun $5\frac{30}{60}$. Consequently, if the diameter of the moon is taken as the unit, the earth's diameter is $3\frac{2}{5}$ of such units, and the sun's diameter $18\frac{4}{5}$. That is to say, the diameter of the earth is $3\frac{2}{5}$ times the diameter of the moon, while the diameter of the sun is $18\frac{4}{5}$ times the diameter of the moon and $5\frac{1}{2}$ times the diameter of the earth.

From these figures the ratios between the solid contents are manifest, since the cube on 1 is 1 unit, the cube on $3\frac{2}{5}$ is very nearly $39\frac{1}{4}$ of the same units, and the cube on $18\frac{4}{5}$ very nearly $6644\frac{1}{2}$, whence we infer that, if the solid magnitude of the moon is taken as a unit, that of the earth contains $39\frac{1}{4}$, and that of the sun $6644\frac{1}{2}$ of such units; therefore the solid magnitude of the sun is very nearly 170 times greater than that of the earth.

The heliocentric system : Copernicus anticipated

ARCHIMEDES, *Psammites (Sand-reckoner)*, c. 1, 1–10.

There are some, King Gelon, who think that the number of the sand is infinite in multitude; and I mean by the sand not only that which exists about Syracuse and the rest of Sicily, but also that which is found in every region, whether inhabited or uninhabited. Again, there are some who, without regarding it as infinite, yet think that no number has been named which is great enough to exceed its multitude. And it is clear that they who hold this view, if they imagined a mass made up of sand as large in size as the mass of the earth, including in it all the seas and the hollows of the earth filled up to a height equal to that of the highest mountain, would be many times further still from recognizing that any number could be expressed which exceeded the multitude of the sand so taken. But I will try to show you, by means of geometrical proofs, which you will be able to follow, that, of the numbers named by me and given in the work which I sent to Zeuxippus, some exceed, not only the number of the mass of sand equal in size to the earth filled up in the way described, but also that of a mass equal in size to the universe. Now you are aware that "universe" is the name given by most astronomers to the sphere the centre of which is the centre of the earth, and the radius of which is equal to the straight line between the centre of the sun and the centre of the earth; this you have seen in the treatises written by astronomers.

But Aristarchus of Samos brought out a book consisting of certain hypotheses, in which the premisses lead to the conclusion that the universe is many times greater than that now so called. His hypotheses are that the fixed stars and the sun remain motionless, that the earth revolves about the sun in the circumference of a circle, the sun lying in the middle of the orbit, and that the sphere of the fixed stars, situated about the same centre as the sun, is so great that the circle in which he supposes the earth to revolve bears such a proportion to the distance of the fixed stars as the centre of the sphere bears to its surface.

Now it is easy to see that this is impossible; for, since the centre of the sphere has no magnitude, we cannot conceive it to bear any ratio whatever to the surface of the sphere. We must, however, take Aristarchus to mean this: since we conceive the earth to be, as it were, the centre of the universe, the ratio which the earth bears to what we describe as the "universe" is the same as the ratio which the sphere containing the circle in which he supposes the earth to revolve bears to the sphere of the fixed stars. For he adapts the proofs of the phenomena to a hypothesis of this kind, and in particular he appears to suppose the size of the sphere in which he represents the earth as moving to be equal to what we call the "universe."

I say then, that, even if a sphere were made up of sand to a size as great as Aristarchus supposes the sphere of the fixed stars to be, I shall still be able to prove that, of the numbers named in the "Principles," some exceed in multitude the number of the sand which is equal in size to the sphere referred to, provided that the following assumptions be made:

1. The perimeter of the earth is about 3,000,000 stades and not greater.[1]

It is true that some have tried, as you are, of course, aware, to prove that the said perimeter is about 300,000 stades. But I go farther and, putting the size of the earth at ten times the size that my predecessors thought it, I suppose its perimeter to be about 3,000,000 stades and not greater.

2. The diameter of the earth is greater than the diameter of the moon, and the diameter of the sun is greater than the diameter of the earth.

In this assumption I follow most of the earlier astronomers.

3. The diameter of the sun is about 30 times the diameter of the moon and not greater.

It is true that, of the earlier astronomers, Eudoxus declared it to be about nine times as great, and Phidias, my father, twelve times, while *Aristarchus tried to prove that the diameter of the sun is greater than* 18 *times, but less than* 20 *times, the diameter of the moon.* But I go even further than Aristarchus, in order that the truth of my proposition may be established beyond dispute, and I suppose the diameter of the sun to be about 30 times that of the moon and not greater.

4. The diameter of the sun is greater than the side of the chiliagon (a regular polygon of 1000 sides) inscribed in the greatest circle in the sphere of the universe.

I make this assumption because Aristarchus discovered that the sun appeared to be about $\frac{1}{720}$th part of the circle of the zodiac, and I myself tried, by a method which I

[1] Archimedes obviously here takes an extreme figure in order that he may be on the safe side.

will now describe, to find experimentally (by means of a mechanical contrivance), the angle subtended by the sun and having its vertex at the eye.

[In the end Archimedes finds, by sheer calculation, that the number of grains of sand that would be contained in a sphere of the size attributed to the universe is less than the number which we should express as 10^{63}.]

PLUTARCH, *De facie in orbe lunae*, c. 6.

Only do not, my good fellow, enter an action against me for impiety in the style of Cleanthes, who thought it was the duty of Greeks to indict Aristarchus of Samos on the charge of impiety for putting in motion the Hearth of the Universe, this being the effect of his attempt to save the phenomena by supposing the heaven to remain at rest, and the earth to revolve in an oblique circle, while it rotates, at the same time, about its own axis.

SEXTUS EMPIRICUS, *Adv. mathematicos*, X, 174.

Yet those who did away with the motion of the universe, and were of opinion that it is the earth which moves, as Aristarchus the mathematician held, are not on that account debarred from having a conception of time.

PLUTARCH, *Plat. quaest.*, VIII, 1, p. 1006 C.

Did Plato put the earth in motion as he did the sun, the moon, and the five planets, which he called the instruments of time, on account of their turnings, and was it necessary to conceive that the earth "rolling about the axis stretched through the universe" was not represented as being held together and at rest, but as turning and revolving, as Aristarchus and Seleucus afterwards maintained that it did, the former stating this as only a hypothesis, the latter as a definite opinion? But Theophrastus adds to his account the detail that Plato in his later years regretted that he had given the earth the middle place in the universe which was not appropriate to it.

ERATOSTHENES

Measurement of the earth

Cleomedes, *De motu circulari corporum caelestium*, I, 10, pp. 90, 20–91, 2; pp. 94, 24–100, 23.

About the size of the earth the physicists, or natural philosophers have held different views, but those of Posidonius and Eratosthenes are preferable to the rest. The latter shows the size of the earth by a geometrical method; the method of Posidonius is simpler. Both lay down certain hypotheses, and, by successive inferences from the hypotheses, arrive at their demonstrations. . . .

The method of Eratosthenes depends on a geometrical argument, and gives the impression of being slightly more difficult to follow. But his statement will be made clear if we premise the following. Let us suppose, in this case too, first, that Syene and Alexandria lie under the same meridian circle; secondly, that the distance between the two cities is 5000 stades; and thirdly, that the rays sent down from different parts of the sun on different parts of the earth are parallel; for this is the hypothesis on which geometers proceed. Fourthly, let us assume that, as proved by the geometers, straight lines falling on parallel straight lines make the alternate angles equal, and fifthly, that the arcs standing on (i.e. subtended by) equal angles are similar, that is, have the same proportion and the same ratio to their proper circles—this, too, being a fact proved by the geometers. Whenever, therefore, arcs of circles stand on equal angles, if any one of these is (say) one-tenth of its proper circle, all the other arcs will be tenth parts of their proper circles.

Any one who has grasped these facts will have no difficulty in understanding the method of Eratosthenes, which is this. Syene and Alexandria lie, he says, under the same meridian circle. Since meridian circles are great circles in the universe, the circles of the earth which lie under them are necessarily also great circles. Thus, of whatever size this method shows the circle on the earth passing through Syene and Alexandria to be, this will be the size of the great circle of the earth. Now Eratosthenes asserts, and it is the fact, that Syene lies under the summer tropic. Whenever, therefore, the sun, being in the Crab at the summer solstice, is exactly in the middle of the heaven, the gnomons (pointers) of sundials necessarily throw no shadows, the position of the sun above them being exactly vertical; and it is said that this is true throughout a space three hundred stades in diameter. But in Alexandria, at the same hour, the pointers of sundials throw shadows, because Alexandria lies further to the north than Syene. The two cities lying under the same meridian great circle, if we draw an arc from the extremity of the shadow to the base of the pointer of the sundial in Alexandria, the arc will be a segment of a great circle in the (hemispherical) bowl of the sundial, since the bowl of the sundial lies under the great circle (of the meridian). If now we conceive straight lines produced from each of the pointers through the earth, they will meet at the centre of the earth. Since then the sundial at Syene is vertically under the sun, if we conceive a straight line coming from the sun to the top of the pointer of the sundial, the line reaching from the sun to the centre of the earth will be one straight line. If now we conceive another straight line drawn upwards from the extremity of the shadow of the pointer of the sundial in Alexandria, through the top of the pointer to the sun, this straight line and the aforesaid straight line will be parallel, since they are straight lines coming through from different parts of the sun to different parts of the earth. On these straight lines, therefore, which are parallel, there falls the straight line drawn from the centre of the earth to the pointer at Alexandria, so that the alternate angles which it makes are equal. One of these angles is that formed at the centre of the earth, at the intersection of the straight lines which were drawn from the sundials to the centre of the earth; the other is at the point of intersection of the top of the pointer at Alexandria and the straight line drawn from the extremity of its shadow to the sun through the point (the top) where it meets the pointer. Now on this latter angle stands the arc carried round from the extremity of the shadow of the pointer to its base, while on the angle at the centre of the earth stands the arc reaching from Syene to Alexandria. But the arcs are similar, since they stand on equal angles. Whatever ratio, therefore, the arc in the bowl of the sundial has to its proper circle, the arc reaching from Syene to Alexandria has that ratio to *its* proper circle. But the arc in the bowl is found to be one-fiftieth of its proper circle. Therefore the distance from Syene to Alexandria must necessarily be one-fiftieth part of the great circle of the earth. And the said distance is 5000 stades; therefore the complete great circle measures 250,000 stades. Such is Eratosthenes' method.

ARATUS

Phaenomena, 1–73.

Let us begin from Zeus, whom we mortals never leave unnamed. Full of Zeus are all streets, all meeting-places of men, full are sea and harbours; every way we stand in need of Zeus. We are even his offspring; he, in his kindness to man, points out things of good omen, rouses the people to labour, calling to their minds the needs of daily life, tells them when the soil is best for the labour of the ox and for the pick, and when the seasons are propitious for planting trees and all manner of seeds. For he it was who fixed the signs in the heaven, and set apart the constellations, and he who, looking to the year, determined which of the stars were best fitted to mark the seasons for men, so that all things might grow unfailingly. For this cause men ever worship him, first and last. Hail, O Father, Great Wonder, great blessing to mankind, hail to thyself and to thy elder line. Hail, ye Muses, most gracious all; and for me who presume to tell of the stars, so far as I rightly may, guide ye all my song.

The stars, many as they are, and scattered hither and thither, are every day ceaselessly borne along in the heaven; but the axis ne'er moves one whit, but remains for ever firmly fixed in the same place, and in the midst it holds the earth in equipoise on all sides, while it carries round the heaven itself.

Two poles form its ends, one each way; one is not seen, while the other faces us to the north, high above the ocean. Encompassing it and together circling round it are the two Bears, which are called Wains. The head of either lies ever towards the flank of the other, and they are ever borne shoulder-wise, their shoulders turned opposite ways. If the tale be true, they were raised to heaven from Crete by the will of Zeus for that, when he was a youth, near the mount of Ida in fragrant Dicton, they hid him in a cave, and nurtured him for a year when the Curetes of Dicton were forswearing Kronos. Now the one men call by the name of Cynosura, the other they call Helice. It is by Helice that the Achaeans on the sea judge where to direct the course of their ships, while the Phoenicians put their trust in the other as they cross the sea. Now Helice is bright and easy to note, appearing large from earliest nightfall; the other is smaller and yet better for sailors, for the whole of it turns in a lesser circuit, and by it the men of Sidon steer the straightest course.

But between them winds, like a branch of a river, that great wonder, the Dragon, bent measureless round about; the Bears are borne on either side of his coil, protected from the dark-blue ocean (i.e. never setting in northern latitudes). The Dragon nears the one with the end of his tail, the other he cuts off all round with his coil. The tip of his tail ends by the head of the Bear hight Helice; Cynosura has her head in his coil; for his coil winds about her very head and comes near her foot, and then turns back and runs upwards. On his head blazes not one star alone but two on his temples and two in his eyes, while one underneath marks the end of his under-jaw. His head lies obliquely, and seems just as if it were nodding towards the end of Helice's tail; his mouth and right temple are quite straight opposite the tip of the tail; the said head moves where the limits of rising and setting are confounded.

Just there wheels a Phantom like to a man toiling painfully. No one knows his story nor to what task he is bound, but all alike call him On his knees.[1] The Phantom that toils on his knees is like one sitting with knees bent; from both shoulders his hands are raised and stretched out, one this way, one that, to a fathom's length. He has the tip of his right foot over the middle of the head of the crooked Dragon.

There, too, that Crown, which the noble Dionysus set to be a memorial of Ariadne dead, wheels beneath the back of the toil-worn Phantom.

[1] The story is, however, contained in the *Catasterismi*, which has come down to us under the name of Eratosthenes, though it can hardly be his in its present form. Here we are told of "Engonasin" that "this, they say, is Heracles, whose foot is on the Serpent (Dragon). He stands clearly visible with his club uplifted, and with the lion's skin wrapped round him. The story is that, when he went in search of the golden apples, he killed the dragon, their appointed guardian, who had been placed there for this very purpose, that he might do battle with Heracles. Hence it was that, when the labour was accomplished at great risk, Zeus, who thought the struggle worthy of a memorial, placed the figure among the stars. The dragon is seen with his head aloft, and Heracles stands over him, having one knee bent but standing with the other foot on the dragon's head, stretching out his right hand with the club in it as if about to strike, and with his left hand holding the lion's skin wrapped about him."

Ib., 91–136.

Behind Helice, like to one driving, comes Arctophylax, whom men call Boötes because he seems to touch the wain-like Bear. He is wholly bright; and below his belt wheels Arcturus himself, a star marked above the rest.

Under the two feet of Boötes you may see the Virgin, who in her hands holds the sparkling Ear of Corn (Spica). Whether she be the daughter of Astraeus whom the story makes the ancient father of the stars, or of someone else, may her course be ever care-free. Another tale is, indeed, current among men, how she once lived upon the earth, and went to and fro before the face of men, and ne'er disowned the races of men or of women of old, but mingled and abode with them, immortal though she was. And they called her Justice; for she gathered together the elders, whether in the market-place or in the wide-spaced street, and, chanting, urged them to decrees favouring the people. They knew naught yet of baneful strife, nor of bickering contention nor uproar, but they led an even life. The cruel sea lay remote, and not yet did ships bring sustenance from afar; but oxen and the plough and Justice herself, queen of peoples, dispenser of things just, supplied all wants abundantly. So long as the earth still nurtured the Golden Race, she abode with them. With the Silver Race she would forgather sparingly, and was no longer wholly at command, for that she yearned for the uses of the ancient race. Yet she still lived on earth in the Silver Age; she would come down from the echoing mountains towards evening, but would no longer consort with any one. But when at times she had filled the great hills with a host of men, she would then threaten them and chide them for their wickedness, declaring that she would never again show herself at their call. "See what manner of men the fathers of the Golden Age have left to follow them, a baser generation! And ye shall breed a baser yet! And there shall, I ween, be wars, there shall be blood shed in enmity among men, and dire suffering shall weigh upon them." With such words she would hie to the hills and leave the people gazing after her still. But when even that generation died, and there followed others, a Race of Bronze, more fell than their forbears, who were the first to forge the death-dealing dagger of the streets, and the first to eat the flesh of the oxen that ploughed, then, indeed, did Justice abhor that race of men, and flew to heaven, and in that place took up her abode where she is seen of men by night, the Virgin, set close to the far-seen Boötes.

HIPPARCHUS, in *Arati et Eudoxi Phaenomena*, I, 1, 3–8.

Several other writers have compiled commentaries upon the *Phaenomena* of Aratus; but the most careful exposition of them is that of Attalus, the mathematician of our own time. Now the explanation of the meaning of the poem I do not regard as requiring great range; for the poet is simple and concise, and also easily to be understood even by readers who are only moderately informed. But to be able to distinguish, in reading what he says about the heavenly bodies, which of his statements are consistent with the observed phenomena and which are erroneous, may well be considered to be a most useful accomplishment, and one specially appropriate to a trained mathematician.

Observing, then, that in very many of the most useful details Aratus is not in agreement with the phenomena which really occur, but that in almost all of these points not only the other commentators but Attalus also agrees with him, I have determined, in view of your enthusiasm for learning, and looking to the benefit of all, to set out the details which seem to me to have been incorrectly given. I proposed to myself this task, not because I was desirous of getting credit for myself out of criticism of others (this would indeed be a vain and ungenerous motive; I hold on the contrary that we should be grateful to all and sundry who undertake laborious personal work for the common benefit of all), but in order that neither you nor any other enthusiasts for learning should fail to get the true view of the phenomena occurring in the universe, as is naturally the case with many persons nowadays; for the charm of poetry invests its content with a certain plausibility, and almost all who expound this particular poet associate themselves with his statements.

Now Eudoxus gives the same collection of phenomena as Aratus, but has set them forth with greater knowledge. It is, therefore, natural that, owing to its agreement with the views of so many distinguished mathematicians, Aratus' poem is accepted as trustworthy. It is, perhaps, not fair to blame Aratus if in some points he is found to be in error; for, in writing his *Phaenomena*, he has followed the arrangement of Eudoxus, without making any observations on his own account, and without professing to be speaking with the authority of a mathematician when giving descriptions of celestial happenings which afterwards prove to be inaccurate.

Ib., I, 2, 1–6.

That Aratus followed Eudoxus' account of the phenomena may be gathered by comparing, at length, Eudoxus' text with that of the poem dealing with the same topic in each case. But a few words may usefully be said here on the subject, because there is uncertainty about it in the minds of most people. Now two books about the phenomena are referred to Eudoxus, and they practically agree in all but a very few details. One of them is entitled the *Mirror*, and the other *Phaenomena*. It is with reference to the *Phaenomena* that Aratus composed his poem.

Thus, in the latter work, Eudoxus writes about the Dragon (Draco) in these terms: "Between the two Bears is the tail of the Dragon, the end-star of which is above the head of the Great Bear. And the Dragon makes a bend by the head of the Little Bear and lies outstretched under its feet; but then, making another bend, it lifts its head up again and puts it forward." Aratus, as it were, paraphrasing this, says:

> "The Dragon nears the one with the end of his tail, the other he cuts off all round with his coil. The tip of his tail ends by the head of the Bear hight Helice; Cynosura has her head in his coil; for his coil winds about her very head and comes near her foot, and then turns back and runs upwards."

And about Boötes Eudoxus says: "Behind the Great Bear is the Bear-minder (Arcturus)," while Aratus has:

> "Behind Helice, like to one driving, comes Arctophylax, whom men call Boötes."

Eudoxus says again: "Under its feet is the Virgin,"

while Aratus has: "Under the two feet of Boötes you may descry the Virgin."

In the case of Engonasin (the Kneeler), Eudoxus says: "Near by the head of the Serpent is the Kneeler with his right foot on the Serpent's head," and Aratus: "He has the tip of his right foot over the middle of the head of the crooked Dragon."

This passage gives the clearest confirmation of what I set out to prove; for both Eudoxus and Aratus are unaware of the truth, which is that the Kneeler has his *left* foot, and not his right, on the head of the Dragon. . . .

HIPPARCHUS, op. cit. I, 4, 1–8.

About the north pole Eudoxus is in error, for he says: "There is a certain star which remains always in the same spot; this star is the pole of the universe," the fact being that at the pole there is no star at all, but there is an empty space, with, however, three stars close to it [probably α and κ of Draco, and β of the Little Bear], with which the point at the pole forms a square, as Pytheas of Massalia also states.

Next, all writers are wrong as regards the position of the Dragon, in that they suppose that his coil is round the head of the Little Bear. For the brightest and leading stars in the quadrilateral of the Little Bear, the more northerly of which is, according to them, at its head, and the more southerly at its forefeet, lie as nearly as possible parallel to the tail of the Dragon. The statement, then, that "Cynosura has her head in his coil; for his coil winds about her very head," is false, though Eudoxus also says the same thing. But Aratus in the case of the Dragon has an error of his own, first in saying that

"The Bears are borne on either side of his coil": for they are on either side of his tail, not his coil. For, as they face opposite ways and, as it were, lie parallel to one another, the tail of the Dragon stretches lengthwise between them, while the coil encloses the Little Bear, but is separated by a long distance from the Great Bear. For the same reason "bent (measureless) round about" is an incorrect description; this would only have happened if the two Bears had been on different sides of the coil.

Again, he is ignorant of the facts when he says:

> "The head lies obliquely, just as if it were nodding towards the end of Helice's tail; his mouth and right temple are quite straight opposite the tip of the tail."

For it is not the right but the left temple which is in a straight line with the tongue of the Serpent (Draco) and the tip of the tail of the Great Bear. To say, as Attalus does, that Aratus supposes the Dragon's head to be turned the contrary way and not towards the interior of the universe, is quite incredible. All the stars have their positions fixed with reference to our point of view, and as if they were turned towards us, save in so far as one or other of them is drawn in profile. Aratus himself makes this clear in many instances; for in all cases where he clearly distinguishes the right or left portion of a constellation his statement agrees with the aforesaid hypothesis. Anyhow, the hypothesis is artistically right and fitting. Attalus, however, will be found to use the same argument to justify also the statement about the left foot of the Kneeler, on which I shall have to speak in the sequel.

Again, with reference to the position of the Dragon's head, Eudoxus' and Aratus' statements agree with the facts, while Attalus' does not. For Aratus, following Eudoxus, says that it moves on the always-visible circle, using these words:

> "Its head moves where the limits of rising and setting are confounded."

But Attalus says that it is slightly more southerly than the always-visible circle, so that it goes below the horizon for a short time. That Attalus is here at variance with the observed fact may be inferred from the following considerations. The star at the front of the Dragon's mouth is distant $34\frac{3}{5}°$ from the pole, his southerly eye is 35°, and his southerly temple 36° from the pole. Now the always-visible circle in the neighbourhood of Athens, where the pointer of the sundial is in length $1\frac{1}{3}$ times its equinoctial shadow, is distant 37° from the pole. Hence, it is clear that the head of the Dragon moves in the always visible part of the heaven, and has only the left temple on the circle itself, and is not, as Attalus says, further to the south, so as to be below the horizon for a short time and then to rise.

POSIDONIUS

Measurement of the earth.

CLEOMEDES, *De motu circulari*, I, 10.

Posidonius says that Rhodes and Alexandria lie under the same meridian. Now meridian circles are circles which are drawn through the poles of the universe, and through the point which is above the head of any individual standing on the earth. The poles are the same for all these circles, but the vertical point is different for different persons. Hence we can draw an infinite number of meridian circles. Now Rhodes and Alexandria lie under the same meridian circle, and the distance between the cities is reputed to be 5000 stades. Suppose this to be the case.

All the meridian circles are among the great circles in the universe, dividing it into two equal parts and being drawn through the poles. With these hypotheses, Posidonius proceeds to divide the zodiac circle, which is equal to the meridian circles, because it also divides the universe into two equal parts, into forty-eight parts, thereby cutting each of the twelfth parts of it (i.e. signs) into four. If, then, the meridian circle through Rhodes and Alexandria is divided into the same number of parts, forty-eight, as the zodiac circle, the segments of it are equal to the aforesaid segments of the zodiac. For, when equal magnitudes are divided into (the same number of) equal parts, the parts of the divided magnitudes must be respectively equal to the parts. This being so, Posidonius goes on to say that the very bright star called Canobus lies to the south, practically on the Rudder of Argo. The said star is not seen at all in Greece; hence Aratus does not even mention it in his *Phaenomena*. But, as you go from north to south, it begins to be visible at Rhodes and, when seen on the horizon there, it sets again immediately as the universe revolves. But when we have sailed the 5000 stades and are at Alexandria, this star, when it is exactly in the middle of the heaven, is found to be at a height above the horizon of one-fourth of a sign, that is, one forty-eighth part of the zodiac circle. It follows, therefore, that the segment of the same meridian circle which lies above the distance between Rhodes and Alexandria is one forty-eighth part of the said circle, because the horizon of the Rhodians is distant from that of the Alexandrians by one forty-eighth of the zodiac circle. Since, then, the part of the earth under this segment is reputed to be 5000 stades, the parts (of the earth) under the other (equal) segments (of the meridian circle) also measure 5000 stades; and thus the great circle of the earth is found to measure 240,000 stades, assuming that from Rhodes to Alexandria is 5000 stades; but, if not, it is in (the same) ratio to the distance. Such then is Posidonius' way of dealing with the size of the earth.

GEMINUS

On physics and astronomy

SIMPLICIUS, in *Phys.* II, 2, 193 b 23 (pp. 291, 21–292, 31, Diels).

ALEXANDER carefully quotes a certain explanation by Geminus taken from his summary of the *Meteorologica* of Posidonius; Geminus' comment, which is inspired by the views of Aristotle, is as follows:

"It is the business of physical inquiry to consider the substance of the heaven and the stars, their force and quality, their coming into being, and their destruction, nay, it is in a position even to prove the facts about their size, shape, and arrangement; astronomy, on the other hand, does not attempt to speak of anything of this kind, but proves the arrangement of the heavenly bodies by considerations based on the view that the heaven is a real Cosmos, and, further, it tells us of the shapes and sizes and distances of the earth, sun, and moon, and of eclipses and conjunctions of the stars, as well as of the quality and extent of their movements. Accordingly, as it is connected with the investigation of quantity, size, and quality of form or shape, it naturally stood in need, in this way, of arithmetic and geometry. The things, then, of which alone astronomy claims to give an account it is able to establish by means of arithmetic and geometry. Now in many cases the astronomer and the physicist will propose to prove the same point, e.g. that the sun is of great size, or that the earth is spherical; but they will not proceed by the same road. The physicist will prove each fact by considerations of essence or substance, of force, of its being better that things should be as they are, or of coming into being and change; the astronomer will prove them by the properties of figures or magnitudes, or by the amount of movement and the time that is appropriate to it. Again, the physicist will, in many cases, reach the cause by looking to creative force; but the astronomer, when he proves facts from external conditions, is not qualified to judge of the cause, as when, for instance, he declares the earth or the stars to be spherical; sometimes he does not even desire to ascertain the cause, as when he discourses about an eclipse; at other times he invents, by way of hypothesis, and states certain expedients by the assumption of which the phenomena will be saved. For example, why do the sun, the moon, and the planets appear to move irregularly? We may answer that, if we assume that their orbits are eccentric circles, or that the stars describe an epicycle, their apparent irregularity will be saved; and it will be necessary to go further, and examine in how many different ways it is possible for these phenomena to be brought about, so that we may bring our theory concerning the planets into agreement with that explanation of the causes which follows an admissible method. *Hence we actually find a certain person*[1] *[Heraclides of Pontus] coming forward and saying that, even on the assumption that the earth moves in a certain way, while the sun is in a certain way at rest, the apparent irregularity with reference to the sun can be saved.* For it is no part of the business of an astronomer to know what is by nature suited to a position of rest, and what sort of bodies are apt to move, but he introduces hypotheses under which some bodies remain fixed, while others move, and then considers to which hypotheses the phenomena actually observed in the heaven will correspond. But he must go to the physicist for his first principles, namely that the movements of the stars are simple, uniform, and ordered, and by means of these principles he will then prove that the rhythmic motion of all alike is in circles, some being turned in parallel circles, others in oblique circles." Such is the account given by Geminus, or Posidonius in Geminus, of the distinction between physics and astronomy, wherein the commentator is inspired by the views of Aristotle.

[1] The theory in question (identical with the Copernican hypothesis) is that of Aristarchus of Samos. The original text obviously had "a certain person (*τις*)" without any name, and "Heraclides of Pontus" was wrongly interpolated.

The Zodiac : motions therein of sun, moon, and planets

GEMINUS, *Elementa astronomiae*, c. 1.

The circle of the signs of the zodiac is divided into twelve parts, and each of the segments has the common name of "a twelfth part," but has also a special name derived from the stars included in it, by which it is given a definite shape, namely Zōidion [a "small picture," or "sign," literally, "small animal"]. The twelve signs are the following: the Ram (Aries), the Bull (Taurus), the Twins (Gemini), the Crab (Cancer), the Lion (Leo), the Virgin (Virgo), the Scales (Libra), the Scorpion (Scorpio), the Archer (Sagittarius), the Horned Goat (Capricornus), the Water-pourer (Aquarius), the Fishes (Pisces).

The word "sign" is used in two senses: (1) for the twelfth part of the zodiac circle, which is a certain distance in space marked off by stars or points; (2) for the picture formed by the stars, according to the resemblance and the position of the said stars.

The "twelfth parts" are equal in size, for the circle of the zodiac is divided by means of the dioptra into twelve equal parts. But the signs made up of fixed stars are neither equal in size nor made up of an equal number of stars; nor do all exactly fill up their own allocated places in the "twelfth parts." Some are deficient and occupy only a small part of the appropriate space, like the Crab; some overlap and encroach on certain portions of the preceding and following signs as, for example, the Virgin. Again, some of the twelve signs do not even lie wholly on the zodiac circle, but some are further north than it is, like the Lion, and some further south, like the Scorpion.

Further, each of the "twelfth parts" is divided into thirty parts, and a single segment is called a "degree," so that the whole circle of the zodiac contains twelve signs and 360 degrees.

The sun traverses the zodiac circle in a year. For a yearly period is that in which the sun travels round the zodiac circle, and returns from the same point to the same point. This period consists of 365¼ days; it is in that number of days that the sun passes the 360 degrees, so that the sun in one day moves over approximately one degree. But a degree is one thing, a day is another. For a degree is a distance which is a thirtieth part of a sign, and a day is a period which is very nearly a thirtieth part of a monthly period. . . .

The yearly period is divided into four parts: spring, summer, autumn, and winter. The spring equinox takes place at the height of flowering time in the first degree of the Ram, and the summer solstice at the period of increasing heat in the first degree of the Crab.

* * * * *

The summer solstice occurs when the sun comes nearest to the region where we live, describing its most northerly circle and producing the longest day of all days in the year, and the shortest night. The longest day is equal to the longest night, and the shortest day to the shortest night. And the longest day contains, in the latitude of Rhodes, 14½ equinoctial hours. The autumn equinox occurs when the sun in its passage from north to south is once more on the equinoctial circle, and makes the day equal to the night. The winter solstice occurs when the sun is furthest away from the place where we live, and is lowest relatively to the horizon, describing its most southerly circle, and producing the longest night of all nights and the shortest day. The longest night contains, in the latitude of Rhodes, 14½ equinoctial hours.

The periods between the solstices and the equinoxes are divided as follows. From the vernal equinox to the summer solstice there are 94½ days. For in this number of days the sun traverses the Ram, the Bull, and the Twins, and, arriving at the first degree of the Crab, brings about the summer solstice. From the summer solstice to the autumnal equinox there are 92½ days, for in this number of days the sun traverses the Crab, the Lion, and the Virgin, and, arriving at the first degree of the Scales, brings about the autumnal equinox. From the autumnal equinox to the winter solstice there are 88⅛ days, for in that number of days the sun traverses the Scales, the Scorpion, and the Archer, and, arriving at the first degree of the Horned Goat, produces the winter solstice. From the winter solstice to the vernal equinox there are 90⅛ days, for in that number of days the sun traverses the remaining three signs, the Horned Goat, the Water-pourer, and the Fishes. The days forming these periods, when all added together, make up 365¼ days, which, as we saw, was the number of days in the year.

At this point the question arises, why, although the four parts of the zodiac circle are equal, the sun, travelling at uniform speed all the time, yet traverses the arcs in unequal times. For the hypothesis underlying the whole of astronomy is that the sun, the moon, and the five planets move at uniform speeds in circles, and in a sense contrary to that of the motion of the universe. The Pythagoreans were the first to approach such questions, and they assumed that the motions of the sun, moon, and planets are circular and uniform. For they could not brook the idea of such disorder in things divine and eternal as that they should move at one time more swiftly, at another time more slowly, and at another time stand still, which last expression refers to what are called the "stations" (or stationary points) in the case of the five planets. No one would credit such irregularity even in the case of a steady and orderly man on a journey. No doubt, the exigencies of daily life are often the cause of slowness and swiftness in men's movements; but when the stars, with their indestructible constitution, are in question, no reason can be assigned for swifter and slower motion.

As regards the remaining heavenly bodies, we shall state the cause elsewhere; here we shall show, in the case of the sun, for what reason, though moving at uniform speed, it nevertheless traverses equal arcs in unequal times.

Above all (in the celestial system) is the so-called sphere of the fixed stars, which includes the imagery of all the signs made up of fixed stars. But we must not suppose that all the stars lie on one surface, but rather that some of them are higher (i.e. more distant) and some lower (less distant); it is only because our sight can only reach to a certain equal distance that the difference in height is imperceptible to us.

Next below the sphere of the fixed stars lies the Shining Star ("Phainon") which goes by the name of Kronos (Saturn). This star traverses the zodiac circle in 30 years, very nearly, and a single sign in 2 years and

6 months. Under the Shining Star, and lower than it, revolves the Bright Star ("Phaëthon"), called the star of Zeus; this traverses the zodiac circle in 12 years, and one sign in one year. Under this is ranged the Fiery Star ("Pyroëis"), that of Ares. This traverses the zodiac circle in 2 years and 6 months, and a sign in 2½ months. The next place is occupied by the sun, which traverses the zodiac circle in a year, and a sign in one month, approximately. Next lower than this lies "Phosphorus" (Lucifer), the star of Aphrodite, and this moves at approximately the same speed as the sun. Below this lies the Gleaming Star ("Stilbon"), the star of Hermes, and it also moves at equal speed with the sun. Lower than all the rest revolves the moon, which traverses the zodiac circle in 27¼ days approximately (the "tropical" month).

If, now the sun had moved on the circle of the fixed signs, the times between the solstices and the equinoxes would have been exactly equal to one another. For, moving at uniform speed it ought, in that case, to have described equal arcs in equal times. Similarly, if the sun had moved in a circle lower than the zodiac circle, but about the same centre as that of the zodiac circle, in that case, too, the periods between the solstices and the equinoxes would have been equal. For all circles described about the same centre are similarly divided by their diameters; therefore, since the zodiac circle is divided into four equal parts by the diameters joining the solstitial and equinoctial points respectively, it would necessarily follow that the sun's circle is divided into four equal parts by the same diameters. The sun, therefore, moving at uniform speed on its own sphere (circle), would in that case have made the times corresponding to the four parts equal. But, as it is, the sun revolves at a lower level than the signs, and moves on an *eccentric* circle, as is explained below. For the sun's circle and the zodiac circle have not the same centre, but the sun's circle is displaced to one side, and, in consequence of its being so placed, the sun's course is divided into four unequal parts. The greatest of the arcs is that which lies under the quadrant of the zodiac circle which stretches from the first degree of the Ram to the 30th degree of the Twins, and the least arc is that which lies under the quadrant from the first degree of the Scales to the 30th of the Archer. . . .

The sun, then, moves at uniform speed throughout, but, because of the eccentricity of the sun's circle, it traverses the quadrants of the zodiac in unequal times.

On day and night

GEMINUS, *Elementa astronomiae*, c. 6.

The word "day" is used in two senses: (1) for the time from the sun's rising to its setting; (2) for the time from the sun's rising to its rising again. The day in the second sense means the revolution of the universe plus the time taken to rise by the arc which the sun describes in its motion in a sense contrary to that of the universe during the time of the revolution of the universe. Hence it is that a day and a night added together are not always equal to another day and night added together. The lengths are equal so far as our sensible perception goes, but, if exactly calculated, they show a small and imperceptible difference. The revolutions of the universe take equal times, but the times taken to rise by the arcs which the sun describes (in its own orbit) during one revolution of the universe are not equal; and it is for this reason that a day and a night added together are not always equal to another day and night added together.

According to the second of the two meanings of the word "day," we say that the month has 30 days, and the year 365¼. A day and a night added together is a period of 24 equinoctial hours, and an equinoctial hour is the 24th part of a day and a night added together.

The lengths of the days are not the same for all countries and cities. The days are longer for those who live towards the north and shorter for those towards the south. The longest day in Rhodes has 14½ equinoctial hours, the longest in Rome 15 equinoctial hours. For those farther north than the Propontis, the longest day has 16 hours, and for those still farther north, 17 and 18 hours.

Now these northern regions are thought to have been visited by Pytheas of Massalia. He says, at all events, in his work "On the Ocean," that: "The barbarians showed us where the sun goes to rest. For it was found that in these regions the night was quite short, consisting in some places of two hours, in others of three, so that only a short interval elapsed from the setting of the sun before it rose again immediately." Crates, the grammarian, observes that even Homer mentions these regions in the passage where Odysseus says:

> "Telepylos of the Laestrygons, where the herdsman driving in his herd calls to herdsman, and the herdsman driving out answers him. There an unsleeping man might earn the wages of two, one

for tending cattle, one for pasturing white-fleeced flocks; for there the paths of night and day come near." (*Odyssey*, X, 82–86.)

For in these regions, the longest day having 23 equinoctial hours, the night is quite short, being left with only one hour, so that the setting of the sun is near to its rising, since only quite a small arc of the summer-tropical circle is cut off under the horizon. If then, says the poet, any one could keep awake through days of this length, he could earn two wages, "one for tending cattle, one for pasturing the white-fleeced sheep." Then he adds the reason, which is mathematical and connected with the theory of a sphere, "for the paths of night and day come near," that is, the time of setting is close to the time of rising.

As we go further northwards, the summer-tropical circle comes to be wholly above the earth, so that at the summer solstice the day there consists of 24 hours. And to those even further to the north a certain part of the zodiac circle is continually above the earth; and those for whom the space of a sign is cut off above the horizon have a day a month long; while for those with whom two signs are cut off above the earth, it is found that the longest day is of two months' duration. And, lastly, there is a place furthest of all to the north, where the pole is vertically overhead, and six signs of the zodiac are cut off above the horizon; for these people the longest day is six months long, and similarly for the night. These regions, also, Homer is thought to mention, according to Crates the grammarian, when he says, about the abode of the Cimmerians:

> "There is the people and state of the Cimmerians, hidden in mist and cloud. Nor does the bright sun ever look down upon them with his rays, either when he mounts to the starry heaven, or when again he turns from heaven to earth; but deadly night is spread over hapless mortals." (*Odyssey*, XI, 14–19.)

For, when the pole is vertically overhead, the effect is to make the night and the day six months long. There are three months in the period during which the sun is passing from the equinoctial circle, which, of course, also occupies the position of the horizon, to the summer-tropical circle, and there is another period of three months, during which the sun comes down again from the summer-tropical circle to the horizon; and all this time the sun will describe parallel circles above the earth. But as it happens that this place of habitation is in the middle of the frigid and uninhabitable zone, the region must all the time be smothered by clouds, and the clouds must be massed together through a great depth of air, so that the rays of the sun are not able to pierce the clouds. Hence it is reasonable that night and darkness should prevail continually in their neighbourhood. For, whenever the sun is above the earth, it is dark with them because of the density of the clouds, and, whenever the sun is under the earth, it is night with them in consequence of the physical necessity, so that their place of habitation is forever unlighted. This, says Crates, is what the poet meant by the words, "Nor does the bright sun ever look down upon them with his rays." Whether this is really what was in the mind of the poet is, let us say, another story. But that there are certain places on the earth, spherical as it is in shape, where the lengths of the days bear to one another the relations aforesaid is clear from the (properties of the) sphere itself. But the places referred to are, in fact, uninhabitable on account of the excessive cold; for they lie in the middle of the frigid zone.

On the other hand, with those who live further south the days are continually shorter and shorter; with some the longest day contains 14 equinoctial hours, with others 13 hours. And, lastly, there is a certain region lying to the south of us which is said to be under the equator, where the poles fall on the horizon, and the sphere of the universe stands straight up, as it were (*sphaera recta*). And all the parallel circles described by the sun in virtue of the revolution of the universe are there bisected <by the horizon>. For this reason there is equinox for ever with the inhabitants of this region.

For the inequality in the lengths of the days is due to no other cause than the elevation of the pole, which is also called the inclination of the universe. . . .

But the increases in the length of days and nights are not equal in all the signs, but in the neighbourhood of the solstitial points they are quite small and imperceptible, so that the length of the days and nights remains the same for about 40 days. For, as the sun approaches and again recedes from the solstitial points, its deviations in latitude are not noticeable, so that it is reasonable that for the aforesaid number of days the sun should appear to our sensible perception to remain in the same position.

Months, years, and cycles

GEMINUS, op. cit., c. 8.

A month is the time from one conjunction to the next conjunction, or from one full moon to the next full moon. A conjunction takes place when the sun and moon are in the same degree, that is, on the 30th day of the moon. "Full moon" means the time when the moon is diametrically opposite the sun, that is, at the "half month." A monthly period consists of $29\frac{1}{2}+\frac{1}{33}$ days. In the period of a month the moon traverses the zodiac circle and in addition the arc by which the sun in the monthly period changes its place in the direct order of the signs, that is, approximately a sign. Hence, in the monthly period, the moon moves approximately through 13 signs.

The exact length of the monthly period is, as we said, $29\frac{1}{2}+\frac{1}{33}$ days, but in ordinary civil life it is taken roughly at $29\frac{1}{2}$ days, so that two months come to 59 days. For this reason the civic months are alternately reckoned as "full" (30 days) and "hollow" (29 days), because two months contain 59 days. Hence the year according to the moon has 354 days. For, if we multiply the $29\frac{1}{2}$ days in a month by 12, the days in the moon-year will make up 354. The moon-year and the solar year are different things, as the solar year is the time of the sun's revolution through 12 signs, that is, $365\frac{1}{4}$ days, while the moon-year contains 12 lunar months, that is, 354 days.

Since, then, neither the month nor the solar year consisted of a whole number of days, the astronomers sought for a period which should contain a whole number of days, a whole number of months, and a whole number of years.

The ancients had before them the problem of reckoning the months by the moon, but the years by the sun. For the legal and oracular prescription that sacrifices should be offered after the manner of their forefathers was interpreted by all Greeks as meaning that they should keep the years in agreement with the sun, and the days and months with the moon. Now reckoning the years according to the sun means performing the same sacrifices to the gods at the same seasons in the year, that is to say, performing the spring sacrifice always in the spring, the summer sacrifice in the summer, and similarly offering the same sacrifices from year to year at the other definite periods of the year when they fell due. For they apprehended that this was welcome and pleasing to the gods. The object in view, then, could not be secured in any other way than by contriving that the solstices and the equinoxes should occur in the same months from year to year. Reckoning the days according to the moon means contriving that the names of the days of the month shall follow the phases of the moon. . . .

Now the ancients reckoned their months at 30 days, and inserted their intercalary months in alternate years. When observation speedily showed this procedure to be inconsistent with the true facts, inasmuch as the days and the months did not agree with the moon, nor the years keep pace with the sun, they sought for a period which should, as regards the years, agree with the sun, and, as regards the months and the days, with the moon, and should contain a whole number of months, a whole number of days, and a whole number of years. The first period they constructed was the period of the *octaëteris* (or eight-year cycle) which contains 99 months, of which three are intercalary, 2922 days, and 8 years. And they constructed it in this way. Since the year according to the sun has 365¼ days, and the year according to the moon 354 days, they took the excess by which the year according to the sun exceeds the year according to the moon. This is 11¼ days. If, then, we reckon the months in the year according to the moon, we shall fall behind by 11¼ days in comparison with the solar year. They inquired, therefore, how many times this number of days must be multiplied in order to complete a whole number of days and a whole number of months. Now the number 11¼ multiplied by 8 makes 90 days, that is, three months. Since, then, we fall behind by 11¼ days in the year in comparison with the sun, it is manifest that in 8 years we shall fall behind by 90 days, that is, three months, in comparison with the sun. Accordingly, in each period of 8 years three intercalary months are reckoned, in order that the deficiency which arises in each year in comparison with the sun may be made good, and so, when 8 years have passed from the beginning of the period, the festivals are again brought into accord with the seasons in the year. When this system is followed, the sacrifices will always be offered to the gods at the same seasons of the year.

They now disposed the intercalary months in such a way as to spread them as nearly as possible evenly. For we must not wait until the divergence from the observed phenomena amounts to a whole month, nor yet must we get a whole month in advance of the sun's course. Accordingly, they decided to introduce the intercalary months in the third, fifth, and eighth years, so that two of the said months were in years following two ordinary years, and only one followed after an interval of one year. But it is a matter of indifference if, while preserving the same disposition of the intercalary months, you put them in other years.

If now it had only been necessary for us to keep in agreement with the solar years, it would have sufficed to use the aforesaid period in order to be in agreement with the phenomena. But, as we must not only reckon the years according to the sun, but also the days and months according to the moon, they considered how this also could be achieved. Thus, the lunar month, accurately measured, having $29\frac{1}{2}+\frac{1}{33}$ days, while the octaëteris contains, with the intercalary months, 99 months in all, they multiplied the $29\frac{1}{2}+\frac{1}{33}$ days of the month by the 99 months; the result is $2923\frac{1}{2}$ days. Therefore in 8 solar years there should be reckoned $2923\frac{1}{2}$ days according to the moon. But the solar year has $365\frac{1}{4}$ days, and 8 solar years contain 2922 days, this being the number of days obtained by multiplying by 8 the number of days in the year. Inasmuch, then, as we found the number of days according to the moon which are contained in the 8 years to be $2923\frac{1}{2}$, we shall, in each octaëteris, fall behind by $1\frac{1}{2}$ days in comparison with the moon. Therefore in 16 years we shall be behind by 3 days in comparison with the moon. It follows that, in each period of 16 years, 3 days have to be added, having regard to the moon's motion, in order that we may reckon the years according to the sun and the months and days according to the moon. But, when this correction is made, another error supervenes. For the

3 days according to the moon which are added in the 16 years give, in ten periods of 16 years, an excess (with reference to the sun) of 30 days, that is to say, a month. Consequently at intervals of 160 years, one of the intercalary months is omitted from <one of> the octaëterides; that is, instead of the three (intercalary) months which fall to be reckoned in the 8 years, only two are actually introduced. Hence, when the month is thus eliminated, we start again in agreement with the moon as regards the months and days, and with the sun as regards the years.

Cycles of Meton and Callippus

Geminus, op. cit., c. 8, 50–60.

Accordingly, as the octaëteris was found to be in all respects incorrect, the astronomers Euctemon, Philippus, and Callippus constructed another period, that of 19 years. For they found by observation that in 19 years there were contained 6940 days and 235 months, including the intercalary months, of which, in the 19 years, there are 7. [According to this reckoning the year comes to have $365\frac{5}{19}$ days.] And of the 235 months they made 110 hollow, and 125 full, so that hollow and full months did not always follow one another alternately, but sometimes there would be two full months in succession. For the natural course of the phenomena in regard to the moon admits of this, whereas there was no such thing in the octaëteris. And they included 110 hollow months in the 235 months for the following reason. As there are 235 months in the 19 years, they began by assuming each of the months to have 30 days; this gives 7050 days.

Thus, when all the months are taken at 30 days, the 7050 days are in excess of the 6940 days; the difference is <110 days>, and accordingly they make 110 months hollow in order to complete, in the 235 months, the 6940 days of the 19-year period. But, in order that the days to be eliminated might be distributed as evenly as possible, they divided the 6940 days by 110; this gives 63 days. It is necessary, therefore, to eliminate the [one] day after intervals of 63 days in this cycle. Thus it is not always the 30th day of the month which is eliminated, but it is the day falling after each interval of 63 days which is called "exairesimos" (to be taken out, eliminable).

In this cycle the months appear to be correctly taken, and the intercalary months to be distributed so as to secure agreement with the phenomena; but the length of the year as taken is not in agreement with the phenomena. For the length of the year is admitted, on the basis of observations extending over many years, to contain $365\frac{1}{4}$ days, whereas the year which is obtained from the 19-year period has $365\frac{5}{19}$ days, which number of days exceeds $365\frac{1}{4}$ by $\frac{1}{76}$th of a day. On this ground Callippus and the astronomers of his school corrected this excess of a <fraction of a> day and constructed the 76-year period out of four periods of 19 years, which contain in all 940 months, including 28 intercalary, and 27759 days. They adopted the same arrangement of the intercalary months. And this period appears to agree the best of all with the observed phenomena.

HIPPARCHUS

Hipparchus' cycle

PTOLEMY, *Syntaxis*, III, 3, Vol. I, p. 207, 7–208, 2, Heib.

Again, in his work "On intercalary months and days," after premising that the length of the year is, according to Meton and Euctemon, $365\frac{1}{4}+\frac{1}{76}$ days, and according to Callippus $365\frac{1}{4}$ days only, he continues in these words: "We find that the number of whole months contained in the 19 years is the same as they make it, but that the year in actual fact contains less by $\frac{1}{300}$th of a day than the odd $\frac{1}{4}$ of a day which they give it, so that in 300 years there is a deficiency, in comparison with Meton's figure, of 5 days, and, in comparison with Callippus' figure, of 1 day." Then, summing up his own views in the course of the enumeration of his own works, he says: "I have also discussed the length of the year in one book, in which I prove that the solar year—that is, the length of time in which the sun passes from a solstice to the same solstice again, or from an equinox to the same equinox—contains $365\frac{1}{4}$ days, less, very nearly, $\frac{1}{300}$th of a day and night, and not the exact $\frac{1}{4}$ which the mathematicians suppose it to have in addition to the said whole number of days."

Discovery of precession of equinoxes

PTOLEMY, *Syntaxis*, VII, 1, Vol. II, p. 2, 11–3, 11, Heib.

First of all we must premise, as regards the name ("fixed stars"), that, since the stars themselves always appear to keep the same figures and the same distances from each other, we may fairly call them "fixed," but, on the other hand, seeing that their whole sphere on which they are carried round as if they had grown upon it, appears itself to have an ordered movement of its own in the direct order of the signs, that is, towards the east, it would not be right to describe the sphere itself also as "fixed." We find both these facts to be as stated, judging by observations made so far as was possible in a comparatively short period. At an even earlier date Hipparchus, in consequence of phenomena which he had recorded, became vaguely aware of the two facts, but, as regards the effects over a longer time, what he gave were guesses rather than facts thoroughly established, because he had come across only very few observations of the fixed stars made before his own time, and, indeed, almost the only observations he found recorded were those of Aristyllus and Timocharis, and even these were neither free from doubt nor thoroughly worked out. We, for our part, have found the same result by comparing observations made to-day with those of the earlier time, but the result is now more firmly established by virtue of the fact that the inquiry has now lasted over a longer period, and the recorded data of Hipparchus about the fixed stars, with which our comparisons have mainly been made, have been handed down to us fully worked out.

Ib., VII, 2, p. 12, 10–13, 9

That the sphere of the fixed stars has a movement of its own in a sense opposite to that of the revolution of the whole universe, that is to say, in the direction which is east of the great circle described through the poles of the equator and the zodiac circle, is made clear to us especially by the fact that the same stars have not kept the same distances from the solstitial and equinoctial points in earlier times and in our time respectively, but, as time goes on, are found to be continually increasing their distance, measured in the eastward direction, from the same points beyond what it was before.

For Hipparchus, in his work "On the displacement of the solstitial and equinoctial points," comparing the eclipses of the moon, on the basis both of accurate observations made in his time, and of those made still earlier by Timocharis, concludes that the distance of Spica from the autumnal equinoctial point, measured in the inverse order of the signs, was in his own time 6°, but in Timocharis' time 8°, nearly. His words at the end are: "If then, for the sake of argument, Spica was, longitudinally with respect to the signs, at the earlier date 8° west of the autumnal-equinoctial point, but is now 6° west of it," and so on. And in the case of practically all the other fixed stars the position of which he has similarly compared he shows that there has been the same amount of progression in the direct order of the signs.

[Ptolemy next (p. 14, 1–15, 5) gives the result of certain observations of his own about the "star in the heart of Leo" (Regulus), showing that, in the interval between Hipparchus' observations and his own, the said star had moved 2° 40′ in the direct order of the signs. As he makes the period 265 years, he observes that the movement has been at the rate of about 1° in 100 years. He adds (p. 15, 17–16, 1):]

"This seems to have been the idea of Hipparchus, to judge by what he says, in his work 'On the length of the year': 'If for this reason the solstices and the equinoxes had changed their position in the inverse order of the signs, in one year, by not less than $\frac{1}{100}$°, their displacement in 300 years should have been not less than 3°.'"

PTOLEMY

The earth does not change its position in any way whatever

Syntaxis, I, c. 7, Vol. I, pp. 21, 9–24, 4, Heib.

In the same way as before it can be proved that the earth cannot make any movement whatever in the aforesaid oblique direction, or ever change its position at all from its place at the centre; for the same results would, in that case, have followed as if it had happened to be placed elsewhere than at the centre. So I, for one, think it is gratuitous for any one to inquire into the causes of the motion towards the centre when once the fact that the earth occupies the middle place in the universe, and that all weights move towards it, is made so patent by the observed phenomena themselves. The ground for this conviction which is readiest to hand, seeing that the earth has been proved to be spherical and situated in the middle of the universe, is this simple fact: in all parts of the earth without exception the tendencies and the motions of bodies which have weight—I mean their own proper motions—always and everywhere operate at right angles to the (tangent) plane drawn evenly through the point of contact where the object falls. That this is so makes it also clear that, if the objects were not stopped by the surface of the earth, they would absolutely reach the centre itself, since the straight line leading to the centre is always at right angles to the tangent-plane to the sphere drawn through the intersection at the point of contact.

All who think it strange that such an immense mass

as that of the earth should neither move itself nor be carried somewhere seem to me to look to their own personal experience, and not to the special character of the universe, and to go wrong through regarding the two things as analogous. They would not, I fancy, think the fact in question to be strange if they could realize that the earth, great as it is, is nevertheless, when compared with the enclosing body, in the relation of a point to that body. For in this way it will seem to be quite possible that a body relatively so small should be dominated and pressed upon with equal and similarly directed force on all sides by the absolutely greatest body formed of like constituents, there being no up and down in the universe any more than one would think of such things in an ordinary sphere. So far as the composite objects in the universe, and their motion on their own account and in their own nature are concerned, those objects which are light, being composed of fine particles, fly towards the outside, that is, towards the circumference, though their impulse seems to be towards what is for individuals "up," because with all of us what is over our heads, and is also called "up," points towards the bounding surface; but all things which are heavy, being composed of denser particles, are carried towards the middle, that is, to the centre, though they seem to fall "down," because, again, with all of us the place at our feet, called "down," itself points towards the centre of the earth, and they naturally settle in a position about the centre, under the action of mutual resistance and pressure which is equal and similar from all directions. Thus it is easy to conceive that the whole solid mass of the earth is of huge size in comparison with the things that are carried down to it, and that the earth remains unaffected by the impact of the quite small weights (falling on it), seeing that these fall from all sides alike, and the earth welcomes, as it were, what falls and joins it. But, of course, if as a whole it had had a common motion, one and the same with that of the weights, it would, as it was carried down, have got ahead of every other falling body, in virtue of its enormous excess of size, and the animals and all separate weights would have been left behind floating on the air, while the earth, for its part, at its great speed, would have fallen completely out of the universe itself. But indeed this sort of suggestion has only to be thought of in order to be seen to be utterly ridiculous.

Arguments against the earth's rotation

Syntaxis, I, 7, Vol. I, p. 24, 5–26, 3, Heib.

Certain thinkers, though they have nothing to oppose to the above arguments, have concocted a scheme which they consider more acceptable, and they think that no evidence can be brought against them if they suggest for the sake of argument that the heaven is motionless, but that the earth rotates about one and the same axis from west to east, completing one revolution approximately every day, or alternatively that both the heaven and the earth have a rotation of a certain amount, whatever it is, about the same axis, as we said, but such as to maintain their *relative* situations.

These persons forget however that, while, so far as appearances in the stellar world are concerned, there might, perhaps, be no objection to this theory in the simpler form, yet, to judge by the conditions affecting ourselves and those in the air about us, such a hypothesis must be seen to be quite ridiculous. Suppose we could concede to them such an unnatural thing as that the most rarefied and lightest things either do not move at all or do not move differently from those of the opposite character—when it is clear as day that things in the air and less rarefied have swifter motions than any bodies of more earthy character—and that (we could further concede that) the densest and heaviest things could have a movement of their own so swift and uniform—when earthy bodies admittedly sometimes do not readily respond even to motion communicated to them by other things—yet they must admit that the rotation of the earth would be more violent than any whatever of the movements which take place about it, if it made in such a short time such a colossal turn back to the same position again, that everything not actually standing on the earth must have seemed to make one and the same movement always in the contrary sense to the earth, and clouds and any of the things that fly or can be thrown could never be seen travelling towards the east, because the earth would always be anticipating them all and forestalling their motion towards the east, insomuch that everything else would seem to recede towards the west and the parts which the earth would be leaving behind it.

For, even if they should maintain that the air is carried round with the earth in the same way and at the same speed, nevertheless the solid bodies in it would always have appeared to be left behind in the motion of the earth and air together, or, even if the solid bodies themselves were, so to speak, attached to the air and carried round with it, they could no longer have appeared either to move forwards or to be left behind, but would always have seemed to stand still, and never, even when flying or being thrown, to make any excursion or change their position, although we so clearly see all these things happening, just as if no slowness or swiftness whatever accrued to them in consequence of the earth not being stationary.

STRABO

On the zones

Geography, II, 2, 1–3; 3, 1.

2, 1. It is one of the things proper to geography to assume that the earth as a whole is spherical in shape, as the universe also is, and to accept all the other inferences which go with this hypothesis.

2. Now Posidonius says that Parmenides was the originator of the division into five zones, but that he represented the "torrid" zone as being about double of its real breadth and overlapping both the tropic circles in the outward direction, that is, towards the temperate zones, whereas Aristotle gives the name "torrid" to the zone between the tropic circles, and calls "temperate" the zones between the tropic and the arctic circles. Posidonius properly raises objections to both these views. He observes that the uninhabitable zone is called "torrid" because of the burning heat, and, if we may base a conjecture on the Aethiopians beyond Egypt, more than half of the breadth of the region between the tropics is not habitable, assuming that the half of the whole breadth is equal to the portion which is cut off by the equinoctial circle in the two directions (north and south) respectively. Now, of the (northern) half, the part from Syene, which marks the summer tropic, to Meroë is 5000 stades; and the distance from Meroë to the parallel of the cinnamon-

producing country is 3000 stades. So far the distances are all measurable, for they are traversed both by sea and by land; but the continuation of the distance as far as the equinoctial circle is found, by calculation based on the measurement of the earth made by Eratosthenes, to be 8800 stades. Therefore, whatever ratio 16800 stades have to 8800, the distance between the tropics will have the same ratio to the breadth of the torrid zone. And if, of the more recent measurements of the earth, we take that which makes it smallest, I mean that of Posidonius, who puts the circumference at 180000 stades, this shows the breadth of the torrid zone to be about half that of the region between the tropic circles, or a little more than half, but by no means equal to, or the same as, that breadth. And how, asks Posidonius, can any one fix the limits of the temperate zones, which are invariable, by means of the arctic circles, which do not exist for all peoples, and are not the same everywhere? That arctic circles do not exist for every one will not affect Posidonius' criticisms, for they must necessarily exist for all inhabitants of the temperate zone, since it is with reference to them alone that the word "temperate" is used. But his point that the arctic circles are not the same everywhere, but are differently situated, is well taken.

3. In making his own division into zones, Posidonius says that five zones are of service with reference to celestial phenomena. Of these, two, namely those under the poles, and stretching as far as the regions which have the tropics as arctic circles, are (he says) "periscian" (i.e. the shadow thrown by an object sometimes goes all round it); the two next to them, stretching as far as the people who live under the tropics, are "heteroscian" (i.e. shadows

fall on one side or the other, north or south, but not both); and the zone between the tropics is "amphiscian" (i.e. shadows fall different ways, north or south, at different times of the year). But, as matter of human interest, there are (according to Posidonius), in addition to these five zones, two other narrow ones, those under the tropic circles, and bisected by those circles, in which the sun is vertically overhead for about half a month in each year. These zones, he says, have a certain peculiarity in that they are exceptionally parched and sandy, and bear nothing but silphium and some pungent fruits withered by heat. For they have no mountains near for clouds to break upon and so produce rain, and no rivers pass through them. Hence it is that they produce creatures with woolly hair, crumpled horns, protruding lips, and flat noses (for their extremities are shrivelled); the fish-eating people, too, live in these zones. That these things are peculiar to these zones is, he says, shown by the fact that those farther to the south have their atmosphere more temperate, and their country is more fruitful and well watered.

Polybius makes six zones; two are those which fall under the arctic circles, two are between the arctic and the tropic circles, and two between these respectively and the equator. However, the division into five zones seems to me to be proper from the point of view of both physics and geography: of physics, because it has reference to celestial phenomena as well as the temperature of the atmosphere, to celestial phenomena because it distinguishes where shadows fall, namely all round, on one side only, or in different directions (north and south) at different times in the year (the best possible way of distinguishing zones), but it also distinguishes conditions for observing the stars which show variations following a certain rough division. . . .

The division into five zones clearly suits geography as well. For it is by one of the temperate zones that geography seeks to define the limits of the section of the earth inhabited by us; this is bounded to the west and to the east by the sea, but to the south and to the north it is defined by the atmosphere, the air in the middle being well-tempered for both plants and animals, while the air on both sides of this is badly-tempered by reason of the excess and defect of heat respectively. . . .

3, 1. When Posidonius adds to the five zones the two zones under the tropics, he does not follow the analogy of the five zones or use a like ground of distinction; he would seem rather to be determining zones by ethnical considerations as well, one of them being the "Aethiopian" zone, another the "Scythian and Celtic," and a third the "intermediate."

2. Polybius again is not right in one respect, that of making certain zones delimited by the arctic circles, two being those which fall under those circles, and two those between the arctic circles and the tropics; for, as we said, we should not define things invariable by means of points that are variable.

TREATISE *DE MUNDO*

De mundo (from Aristotelian corpus), cc. 5–6.

Some have wondered how it is possible that the universe, being constructed out of contrary principles, I mean out of dry and moist, cold and hot, has not long ago been disintegrated and destroyed. Just in the same way some might wonder how a state continues to exist though composed of the most opposite kinds of people, rich and poor, young and old, weakly and robust, bad and good. The persons in question do not realize that this was always the most extraordinary feature of political agreement, that it creates out of many elements one, and out of dissimilar elements a similar, constitution, admitting every sort of character and fortune. Perhaps nature is fond of contraries, and creates the harmonious out of these and not out of like things; just as she mated male with female, and not both with their like respectively, so she created the first concord by bringing together contraries and not likes. And it seems as though art followed nature in doing the same thing, for it is by mixing together the characters of colours, white and black, pale yellow and red, that painting makes its pictures like their originals. Music, too, by mingling high and low, long and short sounds, made harmony between different parts; so also grammar, by making a combination of vowels and consonants, put together its whole system out of these. We find this same idea in Heraclitus "the obscure": "There are matings together of wholes and not-wholes, convergent and divergent, consonant and dissonant; all makes one and one makes all." Thus the constitution of the wholes, I mean of heaven and earth and the whole ordered universe, was disposed in order by one harmony through the blending of the most contrary categories; for by the mixing of dry with moist, hot with cold, light with heavy, and straight with circular, it brought about as an ordered system the whole earth, the sea, the aether, the sun and the whole heaven. One is the force pervading all things, and this, out of things unmixed and different, air, earth, fire, and water, created the whole universe, which it enclosed in one spherical surface, having compelled the most contrary characters in it to agree with one another, and having out of these contrived security for the whole system. The cause of this security is the agreement of the elements, and the cause of this agreement is equal participation and the inability of any one of them to obtain any advantage over another; for light things are in equipoise with heavy, and hot with the contrary, nature teaching us, in the greater issues, that equality is somehow able to preserve concord, and concord to preserve the universe, the begetter of all things, itself possessed of the most perfect beauty. For what created thing is more excellent? Whatever you may allege to be so is a part of it. For everything that is beautiful, everything that is appointed, is called after it, since it is called well "ordered," after the "ordered" universe. And what among particular things can be compared with the arrangement of the heaven and the revolution of the stars and the sun and moon, moving in the exactest measures from age to age? And what infallibility is like that shown by the seasons fair and all-producing, which bring in regular sequence summers and winters, and days and nights, for the due completion of month and year?

In truth, the universe is in size all-supreme, in motion swiftest, in brightness most radiant, in power ageless and deathless. The universe it was that separated the creatures which live in the sea, on land, and in the air, and measured their lives by its own movements. From it all living things draw breath and have soul. Even the strange phenomena which occur in it take place in appointed order, winds of all sorts in confused riot, thunderbolts hurtling from the heaven, and lawless storms breaking loose; the squeezing out of the moisture and the diffusion of fire by these means bring into agreement and stablish the whole. The earth, too, plumed with trees of all kinds, gushing all round with streams, and trodden by living creatures, in season producing, nurturing and harbouring all things, bringing forth countless forms and conditions, keeps its ageless nature unchanged, though shaken by earthquakes, swept by floods, and here and there aflame with conflagrations. And it would seem that all these things coming to it from the good God assure its safety for ever and ever. For, when it is shaken by an earthquake, the gases that have filtered into the earth escape, finding vent at the breaches made in it; purified by showers, it is washed clean of all disorders; blown upon by breezes all round, it is cleansed in the parts under it, and in the parts above it. Further, the flames soften the frost-bound, the frost abates the flames. Of particular things, some are ever coming to birth, some are at their prime, some passing away; the births balance the wastings, and the wastings lighten the births. And the one state of security to which all things contribute through their being ceaselessly pitted one against the other, now mastering and now being mastered, keeps the whole indestructible for ever.

It remains to speak summarily, in the manner in which we discussed the other topics, of the cause which keeps the whole together; for it would be wrong, in speaking about the universe, albeit not in precision of detail, but yet in such a way as to give a general idea, to pass over the most sovereign power in the universe. It is an ancient saying, and one traditional among all men, that all things come to us from God and are caused by God, and nothing in nature is, in itself, self-sufficient if bereft of the security proceeding from him. Hence some of the ancients were prompted to declare that all these things are full of gods, everything that appears to us through the medium of sight, hearing, and all our senses; and the account given by these thinkers is a fitting tribute to the divine power, but not to the divine essence. For God is indeed the preserver of all things and the author of all things whatsoever which are accomplished in this world, yet herein he does not endure the fatigue of a living creature working with his own hands and toiling, but he only applies the untiring power through which he holds sway over things seemingly remote. He himself has the uppermost and chiefest seat, and for this reason has received the name of the Highest, enthroned, as the poet says, in the "topmost pinnacle" of the whole heaven; that body which is continually nearer to him profits, so to say, the most by his power, then the body next in succession to it, and so on in order down to the regions where we live. Wherefore the earth and all that is upon it, being at the greatest distance from the benefit that comes from God, are feeble and ill-matched, and full of endless disorder; not but what, so far as it is the nature of the divine to penetrate everywhere, this occurs equally in our world and in the regions above us, all of which share in the divine beneficence more or less fully, according as they are nearer to or further away from God. It is, therefore, better to embrace the view which is becoming, and most befits God, namely that the power seated in heaven is, as we may say, the cause of security to one and all, even of those who are the furthest away from him, rather than to suppose that it passes through, and sojourns in, the place where it is not meet or right for it to go, and there directly governs the things on the earth. It is not even appropriate work for leaders of men to supervise any and every duty; thus it would not be proper that the commander of an army or the head of a state or household should be called upon to tie up bed-clothes in a sack, or to do any menial task, such as could be done by any slave; but we should take example by the story of the Great King. The pomp displayed by Cambyses, Xerxes, and Darius was organized in the grand style, aiming at the highest degree of solemnity and loftiness. He himself, as the story goes, had his abode at Susa or Ecbatana, where, unseen of any one, he occupied a wonderful palace, with an enclosure round it, gleaming with gold, silver-gilt, and ivory; gateways there were many and close together; porches many stades apart from one another; these were fortified with bronze doors and great walls. In addition, there was an ordered array of the foremost and noblest men, some about the person of the king, bodyguards, and attendants, others as guards of each enclosure, the so-called "warders" and "listeners," as though the king in person, called master and god, should see everything and hear everything. Apart from these, there were other established offices of stewards of revenues, leaders in war and of the chase, receivers of gifts, and others in charge, severally, of the remaining functions as circumstances required. The whole empire of Asia bounded by the Hellespont on the west and the Indus on the east was distributed according to tribes among generals and satraps and kings, servants of the Great King, to whom, again, were subordinated day-couriers, watchmen, messengers, guards, and overseers of signal-stations. Such was the organization, and most important of all were the signal-stations flashing signals to one another in succession from the limits of the empire as far as Susa and Ecbatana, so that the king heard, day by day, news of all happenings in Asia. Now we must regard the supremacy of the Great King as inferior to that of the God who rules the world in the same degree as the state of the meanest and feeblest of living creatures falls short of that of the Great King. Hence, if it was accounted unworthy that Xerxes should be thought to do everything himself, both giving such orders as he pleased and supervising administration, it would be much more unbecoming to God. The worthiest and most proper view is that he is enthroned in the highest place, and that his power, pervading the whole world, moves the sun and the moon, and wheels round the whole heaven, and is the cause of security to all that is on the earth. No need has he of machinery or of the service of others, as those who rule among us need the work of many hands on account of their weakness; this was ever the supreme attribute of the divine to be able to create forms of all kinds without effort and by mere motion, just as, for instance, engineers by one movement of a machine can perform many and various operations. In like manner, puppet-showmen, by pulling one string, make a puppet-creature move neck and hand and shoulder and eye, and sometimes every part, with a certain rhythmic motion. Thus it is that the divine nature, starting with a certain simple motion of that which is first in order, communicates its power to the things contiguous to it, and from them again to what is more remote, until it pervades the whole. For one thing is moved by another, and in its turn sets another in motion, in due order, all acting in a manner appropriate to their own construction, their methods being not all the same but different and various, nay, sometimes opposed, although the first keynote, so to speak, struck to start the motion is always one. It is just as if one were simultaneously to throw out of a vessel a sphere and a cube, and a cone and a cylinder—each of them will move in a manner corresponding to its own particular shape—so, too, if one had in the folds of a garment a water animal, a land animal, and a winged creature, and let them all go at once; for it is clear that the swimming creature will take a leap into its own proper abode and swim away, the land-animal will creep away to its own haunts and feeding places, while the denizen of the air will rise aloft from the earth and fly away, the one first cause having assigned to each its own means of well-being. Thus it is with the universe also; for by means of the simple revolution of the whole heaven, marked off by day and night, the different courses of all the heavenly bodies are produced, though they are all enclosed by one sphere, and some move more swiftly and others more slowly, according to the distances between them, and their own constitutions, respectively. For the moon completes its own circle in a month, waxing and waning and vanishing, the sun and the stars which accompany it in its course, the so-called Morning Star, and the star of Hermes, in a year, the Fiery Star in double of that period, the star of Zeus in six times that again (i.e. twelve years), and last of all, the star of Kronos, so-called, in two and a half times the period of the star next below it. One musical scale arising from all these as they join in chorus and dance in the heaven is born of unity and ends in unity, and has given truthfully to the whole the name of Cosmos, "ordered universe," not disorderliness. Just as in a chorus, when the leader has struck up, the whole chorus of men, sometimes of women too, joins in, and the combination of different voices higher and lower in pitch makes one well-proportioned harmony, so it is with God who sways the whole universe; for according to the keynote given from on high by him who might, by analogy, be called the chorus-leader, the stars and the whole heaven forever move. The all-lighting sun travels two ways; in one of these he delimits the day and night by his rising and setting, while in the other he brings the four seasons of the year, as he diverges forward to the north and backward to the south. In due season come rains and winds and dews, and the different conditions which arise in the region enveloping the earth, thanks to the first cause. Next to these come the gushings-forth of rivers, risings of the sea, growths of trees, ripenings of fruits, births of living creatures, nurturings, maturings, and decayings of all things, to which, as I said, the construction of each individual thing contributes. When, therefore, the sovereign and father of all things, invisible to anything but reason, gives the signal to every natural body that ranges between heaven and earth, each moves continually in its own circle and within its own limits, now disappearing from view, now appearing, revealing countless forms, and again hiding them, at the bidding of one principle. What is done seems very like what happens especially in times of war, when the trumpet gives the signal to the army; for then every one hears the sound, and one takes up his shield, one his greaves, his helmet or his baldric, and one man bridles his horse, one mounts his chariot, and another passes on the watchword; forthwith the captain of a company joins his company, the commander of a squadron his squadron, the horseman his troop, the light-armed soldier hurries to his place, and all move to obey one giver of orders, according to the dispositions of the officer in supreme command. We must think in this wise about the universe also; for, all being prompted by one tendency, the action proper to each is taken, though the tendency itself is invisible and imperceptible, which fact, however, is no obstacle to its operation or our obedience to it; for the soul, by reason of which we live and have our houses and states, is itself invisible, but is seen by its works. For the whole ordering of life was by it discovered and disposed, and is by it kept together, ploughings of the ground and plantings, inventions in crafts, observance of laws, ordering of the constitution, internal administration, war outside the frontiers, and peace. We must so think about God also, who is most mighty in power, most fair in beauty, immortal in life, most excellent in virtue; for, though unrevealed to any mortal being, he is revealed in his actual works. For all experiences, and all things in the air, on the earth, and in the water may truly be called the works of God, the ruler of the universe, from whom, in the words of Empedocles, the philosopher of nature,

> "all things had their birth, all things that were, all that are, and all that shall be, trees, men and women, beasts, birds of prey, and fishes that are nurtured in the water."

It is truly right, though the illustration is a humble one, to compare the world to the so-called keystones in barrel-vaults, which, lying in the midst, by virtue of their connexion with both sides, keep the whole figure of the vault in harmony, in position, and unmoved. There is a story, too, of the sculptor Phidias, how that, when setting up the Athena on the Acropolis, he modelled in the middle of her shield his own face, and by some hidden device so fastened it to the statue that, if any one wished to take it off, he would be forced to break up and destroy the whole statue. Such is the relation of God in the universe, since he maintains the harmony and security of the whole, though he is not at the centre, where lie the earth and all this troubled region, but stands above, all pure, in a pure region which we correctly call heaven (*ouranos*) from its being the boundary (*hŏros*) of the things above (*anō*), and Olympus as being wholly bright and divorced from all gloom and all irregular movement, such as arises among us through storms and violent winds, as the poet Homer says, in the words:

> "To Olympus where, they say, is the seat of the Gods, for ever secure; for neither is it shaken by winds, nor ever is wet with rain, nor does snow come near it, but clear and cloudless sky is outspread and the white light of day shoots up."

CLEOMEDES

On a "paradoxical" eclipse of the moon

De motu circulari, II, 6, pp. 218, 8–226, 6.

These facts having been proved with regard to the moon, the argument establishing that the moon suffers eclipse through falling into the earth's shadow would seem to be contradicted by the stories told about a class of eclipses seemingly paradoxical. For some say that an eclipse (sometimes) occurs, even when both the luminaries are seen above the horizon. This should make it clear that (in that case) the moon does not suffer eclipse through falling into the earth's shadow, but in some other way, since, if an eclipse occurs when both sun and moon appear above the horizon, the moon cannot suffer eclipse through falling into the earth's shadow. For the place where the moon is, when both bodies appear above the horizon, is still being lit up by the sun, and the shadow cannot yet be at the place where the moon gives the impression of being eclipsed. Accordingly, if this be the case, we shall be obliged to declare that the cause of the eclipse of the moon is a different one. Such being the story, the more ancient of the mathematicians tried to get rid of the difficulty in this way. They argued that it is not impossible, even when both luminaries are above the horizon, for the moon to fall into the earth's shadow and to be exactly opposite to the sun. On the assumption that the shape of the earth is flat and plane, this could not happen; but, seeing that the figure formed by it is spherical, it would not be impossible that the two divine bodies should be seen above the horizon, while being exactly opposite each other. They will not, it is true, be in sight of one another while diametrically opposite to one another, because of the prominences formed by the convexities on the earth's surface; but persons standing on the earth would not be prevented from seeing them both, provided they stood on the convexities of the earth, which are no obstacle to those standing thereon seeing both bodies above the horizon, though the convexities do intervene between the bodies diametrically opposite to one another. The bodies will not be in view of one another; but we shall not be prevented from seeing them both if we are standing on the convexities of the earth, which intervene between the bodies themselves, being, as they are, in the depressions about the horizon, while the convexities on which we stand are more lofty.

Such is the solution which the more ancient of the mathematicians give of the difficulty alleged. But we may feel doubt of the soundness of the line taken by them. For, if our eye were situated on a height, the effect might be as stated, if, I mean, we were raised far away from the earth into the air, but it could not possibly happen if we stood on the earth. For, though there may be some convexity on which we stand, our sight itself becomes evanescent owing to the size of the earth. Hence we should altogether refuse to admit or believe that the thing is possible, I mean that an eclipse of the moon can occur when both bodies are seen above the horizon by us standing on the earth and at a lower level.

First, we must take a fundamental objection, and maintain that this story has been invented by some persons who desired to cause perplexity to the astronomers and philosophers who concern themselves with these things. For, while there have been many eclipses, both total and partial, and all have been recorded, history knows, at all events down to our own day, of no person having noted any eclipse of this sort, no Chaldaean, no Egyptian, no other mathematician or philosopher; nay, the story is pure fiction. Secondly, if the moon had suffered eclipse in any other way, and not through falling into the earth's shadow, it would also at times have suffered eclipse when it was not full moon, and when it was, more or less, in advance of the sun, and again when, after the full moon, it again approached the sun and waned. But, as it is, although very many eclipses of the moon have occurred (for the eclipse is not even a rare phenomenon), it has never suffered eclipse without being full and without being diametrically opposite to the sun, and it has only been eclipsed when it was possible for it to come into the earth's shadow. Moreover, nowadays all lunar eclipses are foretold by authorities on the Canon, because they know that, whenever it occurs, the moon is found to be full and to be, either in whole or in part, directly under the midmost circle of the zodiac, thus making the eclipses partial or total, as the case may be. It is, therefore, impossible for eclipses of the moon to occur when both luminaries are seen above the horizon.

Nevertheless, having regard to the many and infinitely various conditions which naturally arise in the air, it would not be impossible that, when the sun has just set, and is under the horizon, we should receive the impression of its not yet having set, if there were cloud of considerable density at the place of setting and the cloud were illuminated by the sun's rays and transmitted to us an image of the sun, or if there were "anthelium." Such images are indeed often seen in the air, especially in the neighbourhood of Pontus. The ray, therefore, proceeding from the eye and meeting the air in a moist and damp condition might be bent, and so might catch the sun although just hidden by the horizon. Even in ordinary life we have observed something similar. For, if a gold ring be thrown into a drinking cup or other vessel, then, when the vessel is empty, the object is not visible at a certain suitable distance, since the visual current goes right on in a straight line as it touches the brim of the vessel. But, when the vessel has been filled with water up to the level of the brim, the ring placed in the vessel is now, at the same distance, visible, since the visual current no longer passes straight on past the brim as before, but, as it touches, at the brim, the water which fills the vessel up to the brim, it is thereby bent, and so, passing to the bottom of the vessel, finds the ring there. Something similar, then, might possibly happen in a moist and thoroughly wet condition of the air, namely that the visual ray should, by being bent, take a direction below the horizon, and there catch the sun just after its setting, and so receive the impression of the sun's being above the horizon. Perhaps, also some other cause akin to this might sometimes give us the impression of the two bodies being above the horizon, though the sun had already set. But the observed phenomena make it as clear as day that the moon is not eclipsed otherwise than by falling within the earth's shadow. So much for eclipses.

PLUTARCH

On the face in the moon

De facie in orbe lunae, cc. 5–10, 16, 21–22.

[In the first four chapters of the dialogue, as we have it, two suggestions about the "apparent face in the moon," are disposed of: (1) that the appearance is simply an optical illusion due to defects in our vision; (2) that it is a reflection, as in a mirror, of the Great Sea or Ocean. With chapter five begins a discussion of other hypotheses.]

5. "Do not," said Lucius, "let it be thought that we

are simply pouring insult on Pharnaces by thus passing over the Stoic view without a word addressed to it; do therefore, by all means, make some reply to the man who supposes that the moon is a mixture of air and of a mild form of fire, and then asserts that, just as a ripple spreads over calm water, so blackening air creates the impression of a certain figure (on the surface of the moon)."

"It is good of you, Lucius," said I, "to cloak the absurdity of this view in such fair-sounding terms; this was not so with our friend, who used to say, what was true, that these thinkers gave the moon 'a black eye' when they filled her face with spots and black patches, calling her, at the same time, Artemis and Athena, and yet making her out to be a combination and hotch-potch of dusky air and fire like charcoal, with no power of kindling anything, and with no light of her own, a sort of body hard to define, for ever smouldering and being baked, like those thunderbolts which the poets describe as 'lustreless' and 'sulphurous.' But that a charcoal fire such as they make the substance of the moon to be cannot have any permanence or consistency at all, unless it has a supply of solid fuel to sustain and feed it, is a truth which it would seem that certain philosophers are less capable of appreciating than would be the people who say in jest that Hephaestus has been represented as lame because fire cannot any more go on without wood than a lame man without a stick! If the moon is fire, whence has such a quantity of air come to be inside it? This region above, which is carried round in a circular motion, does not consist of air, but of a superior substance which is by nature capable of rarefying and helping to kindle all things. And, if the air has come to be in it, how is it that it has not gone away, taking some other form after being etherealized by fire, but instead keeps its character and dwells with the fire such a long time, as if it were a nail fitted always to the same parts and tightly riveted? Had it been rare and diffused, it should not have remained where it was, but should have been dissipated; it could not have become solid if it had been mixed with fire, and had in its composition no moisture or earth, for it is these things alone which are naturally capable of solidifying air. Moreover, rapid motion actually fires the air contained in stones, and even in cold lead; much more, then, should the air in fire be fired when whirled round at such a speed. They quarrel with Empedocles, who made the moon a mass of frozen air like hail surrounded by a sphere of fire; and yet they themselves say that the moon, being of fire, encloses air within her, scattered, some here, some there, though she has in her no cracks nor holes nor hollows such as are conceded to her by those who make her earthy in nature, and therefore she must clearly have the air resting on her convex surface. This is absurd from the point of view of permanence, and is impossible having regard to what we see at full moon; for there ought not, then, to have been any distinction of black and shadowy, but either all ought to have been equally dimmed, i.e. when covered, or all should have been bright together, when, namely, the moon was caught by the sun. Even with us, the air in the deep and hollow places of the earth, where light does not penetrate, remains shady and unilluminated, whereas that which is outside and diffused about the earth has light and bright colour; for its rarity makes it easily mix with any quality or character whatever, and in particular, if it so much as grazes, as you say, or touches, light, it is transformed and illuminated throughout. This same fact seems entirely to support the theory of those who thrust the air in the moon into deep places and ravines, but is a fatal refutation of you who, I know not how, make a mixture of air and fire and compose the sphere out of them. For it is impossible that shadow should remain on the surface of the moon when the sun shines with its light on the whole of that section of the moon which is visible to our sight."

6. While I was still speaking, Pharnaces broke in: "Here, again, we have employed against us the stock device borrowed from the Academy, that of taking care, every time that they discuss things with others, not to allow their own opinions to be criticized, but always to put the others, whenever they meet them, in the position of defendants, not accusers. But you will not to-day draw me into defending the views you impute to the Stoics before you have rendered an account of your own action in turning the universe upside down." Lucius smiled and said: "Very well; only do not bring against me a charge of impiety such as Cleanthes used to say that it behoved Greeks to bring against Aristarchus of Samos for moving the Hearth of the Universe, because he tried to save the phenomena by the assumption that the heaven is at rest, but that the earth revolves in an oblique orbit, while also rotating about its own axis. Now we put forward no view of our own; but how do those who assume that the moon is an earth turn things upside down any more than you do when you place the earth where it is, suspended in the air, though it is much larger than the moon, as the mathematicians have shown by

measuring the size of the obstruction at the time of an eclipse and the passage of the moon through the earth's shadow? For the shadow of the earth is less as it extends, since the illuminating body is greater than it, and the fact that the shadow is thin and narrow at the end did not, so they say, escape Homer, but he called the night 'swift' (or 'pointed') because of the shadow coming to a sharp point. Anyhow, by means of the fact stated, the moon's size is detected at eclipses when she barely emerges from the shadow in three times her own diameter. Think how many moons must go to the earth when it makes a shadow which at the narrowest part (traversed by the moon) is equal to three moons. But yet you fear that the moon may fall. Perhaps in this you are affected by what Aeschylus says of the earth, how Atlas 'stands firm, bearing up on his shoulders the pillar of heaven and earth, a burden not easy for the arms.' Your fear is that only light air circulates below the moon, and that it is not adequate to bear a solid weight, whereas the earth is, in the words of Pindar, girt about 'with pillars based on adamant.' This is why Pharnaces is not for himself apprehensive of the earth's falling, but pities those situated under the orbit of the moon, Aethiopians or Taprobanes, say, for being liable to have a weight of that size fall on them. Yet the moon has a security against falling in her very motion and the swing of her revolution, just as objects put in slings are prevented from falling by the circular whirl; for everything is carried along by the motion natural to it if it is not deflected by anything else.[1] Thus the moon is not carried down by her weight, because her natural tendency is frustrated by the revolution. Nay,

[1] This is practically Newton's first Law of Motion.

there would, perhaps, be more cause for wonder if she were altogether at rest and unmoved like the earth. As it is, the moon has great reason not to fall upon us, whereas the earth, having no part in any other movement, might have been expected to move under the influence of its weight acting alone; and it is heavier than the moon, not only in proportion to its size, but even in greater proportion, since the moon has become light through heat and burning. Altogether, it would seem from what you say, that, if she is fire, the moon has all the more need of earth, and matter on which she stands and to which she clings, so holding together and keeping alight her force, for it is not possible to conceive of fire being conserved without fuel. But you maintain that the earth remains fixed without base or root." "Yes, of course," said Pharnaces, "it keeps its central place as being proper and natural to it; for it is this place about which weights with their natural tendency press forward and move, converging from all directions; whereas all the upper region, even if it receives anything of an earthy nature thrown up into it, straightway thrusts it out hitherward, or rather drops it, to be carried down by its own natural tendency."

7. At this point, wishing to give Lucius time to refresh his memory, I called Theon and said, "Which of the tragic poets is it, Theon, who has said that 'physicians purge bitter bile with bitter drugs'?" and, on his answering that it was Sophocles, I said, "We are forced to concede this to physicians, but we must not listen to philosophers, if they claim to meet paradoxes with paradoxes, and controvert surprising doctrines by inventing others still more strange and surprising, as these people do with

their idea of motion towards the centre. What absurdity is there that this does not imply? Does it not mean that the earth is a sphere, though it contains such enormous depths and heights and irregularities? That people dwell at our antipodes, like wood-worms or lizards, clinging to the earth with their lower limbs upwards? That we ourselves do not remain perpendicular as we walk, but remain at an angle and sway like drunken men? That masses weighing a thousand talents, borne down through the depth of the earth, come to rest when they reach the centre, though nothing meets or resists them; and that, if in their downward rush they should overshoot the centre, they would turn back again and reverse their course of themselves; that segments of beams, when they are sawn through at the surface of the earth on either side, do not move downwards throughout, but, as they fall on the earth, receive a thrust from outside inwards, and are lost about the centre; that a rushing stream of water falling downwards, if it came to the centre, which they themselves declare to be incorporeal, would halt suspended round it, or circle about it, oscillating with an oscillation which never stops and never can be stopped? Some of these things a man might, without perjuring himself, force himself to represent as imaginable by his intelligence. But this is making up down, and everything topsy-turvy, with a vengeance, if things from us to the centre are 'down,' and things under the centre are 'up' again. This would mean, for instance, that, if a man through sympathy with the earth should stand with the centre of the earth at his middle, he would have his head up and his feet up too; that, if a man dug into the part beyond the centre, he would have his head up but the stuff dug out would be pulled from above downwards; and that, if another man could be conceived as standing the reverse way to him, the feet of both alike would be 'up,' and would be called so.

8. "Such and so many are the paradoxes of which they have shouldered and trail along—not, by Zeus, a mere bag-full, but some juggler's whole stock-in-trade and show-room; and then they say that others are talking nonsense in placing the moon, being an earth, up aloft, and not where the centre is. Yet, if all heavy bodies converge to one and the same point, while each presses on its own centre with all its parts, it will not be so much *qua* centre of the universe as *qua* whole, that the earth will appropriate weights, because they are parts of itself; and the tendency of bodies will be a testimony, not to the earth of its being the centre of the universe, but, to things which have been thrown away from the earth and then come back to it, of their having a certain community and natural kinship with the earth. . . .

"But, if any body has not been allotted to earth from the beginning, and has not been rent from it, but somehow has a constitution and nature of its own, as they would maintain to be the case with the moon, what is there to prevent its existing separately and remaining self-contained, compacted and fettered by its own parts? For not only is the earth not proved to be the centre, but the way in which things here press and come together towards the earth suggests the manner in which it is probable that things have fallen on the moon, where she is, and remain there. Moreover, I do not see why the thinkers who drive earthy and heavy things together into one place and make them parts of one body, do not also put light things under similar compulsion, but allow such immense structures of fire to remain separate; why do they not gather all the stars into one place and express a clear conviction that all upward-tending and flame-like objects ought to make up one common body?"

9. "But," said I, "my dear Apollonides, you say that the sun is distant from the outermost revolving sphere by countless myriads of stades, and that the Morning Star next to it, then the 'Gleaming' Star ('Stilbon, i.e. Mercury) and the other planets, revolve below the fixed stars and at great distances from one another, but you suppose that the universe provides within itself no free space or interval for heavy and earthy bodies. You must see that it is ridiculous that we should deny that the moon is earth because she is removed from the lower region, and yet declare that she is a star, when we see that she is banished from the outermost revolution by so many myriads of stades and, as it were, sunk into a sort of abyss; at all events she is so much lower than the fixed stars that no one can state the measure of the distance, since numbers do not suffice for you mathematicians when you try to calculate it, but she in a way touches the earth and, in her revolution near to it, 'rolls like the nave of a chariot-wheel,' as Empedocles has it, 'round the furthest (goal).'

"Nor does she often clear the earth's shadow, since she rises so little, though the illuminating body (the sun) is so enormous, but she seems to wander so close to the surface, and so to speak in the embrace, of the earth that she is shut off from the sun's light by it, because she does not rise above this shadowy, dark, terrestrial region, which is the inheritance of earth. Therefore I think we must have the courage to declare that the moon is within the confines of the earth, since she is obscured by the earth's extremities.

10. "Leaving out of consideration the other stars, both the fixed stars and the planets, consider what Aristarchus proves in his treatise On Sizes and Distances, namely that the distance of the sun from us is more than 18 times, but less than 20 times, the moon's distance from us. But the authority who places the moon highest says that she is distant from us 56 times the earth's radius, which latter is, on a moderate estimate, 40000 stades; on this basis the distance of the sun from the moon works out at more than 4030 myriads of stades. So far is she placed away from the sun on account of her weight, and so near does she approach the earth, that, if you must divide estates according to localities, the inheritance of the earth and her beauty appeal to the moon, and the moon has the next title to the goods and persons on the earth in virtue of her kinship and proximity. And in my opinion we do no wrong if, while we assign such depth and dimensions to what we call the upper regions, we reserve to what is below a space for circulation, with breadth corresponding to the distance from the earth to the moon. For the person who restricts the term 'upper' to the outermost surface of the heaven and calls all the rest 'lower,' goes too far, and on the other hand, he who limits 'lower' to the earth, or rather its centre, is outside the pale . . ."

16. . . . (Lucius to Aristotle.) "As to the other stars, and the whole heaven, when your school (the Peripatetic) gives them a character which is pure and transparent and exempt from all change due to passion,

and when you carry them in a circle of eternal and never-ending revolution, we might perhaps not quarrel with you, on the present occasion at least, though there are countless difficulties. But, when the theory comes down and touches the moon, she no longer keeps her impassiveness and that beauty of form which the others have. To pass over the other inequalities and differences, this same face which appears in the moon has come there through some affection of her substance or some intermixture with another substance. Now that which is mixed with something else is affected by it; it loses its purity when forcibly filled with matter inferior to itself. And in regard to the moon's sluggishness and slowness of speed, and her faint and imperceptible heat by which, as Ion says, 'no grape is ripened black,' to what shall we attribute them but to her own weakness and passiveness, if passiveness can attach to an eternal and celestial body? For, to put it shortly, my dear Aristotle, as an earth she seems a perfectly beautiful, noble, and well-ordered thing, but, as a star or luminary or a divine and heavenly body, I fear she will prove unshapely and uncomely, and will do no credit to her beautiful name—if, I mean, of all the vast multitude of bodies in the heaven, she alone wanders about begging light from others, 'ever wistfully gazing at the sun's rays,' as Parmenides says. Our friend, on bringing up in the conversation the dictum of Anaxagoras that 'the sun puts the brightness in the moon,' was applauded by the company; however, I will not repeat what I learned from you or with you, but I will gladly pass to the remaining points. It is probable that the moon is lit up, not as crystal or ice would be, by the sun shining in and through her, nor yet by the collection of light and rays, as torches multiply their light. Were this the case, we should have full moon just the same at the new moon or at the middle of the month—that is, if the moon does not hide or block out the sun, but the sun's light either passes through her on account of her rarity, or shines into her by way of intermixture, and helps in kindling the light about her. For it is not possible, at the time of conjunction, to blame any turning aside or swerving on her part, as when she is halved or doubly convex or crescent-shaped; nay, being then placed 'vertically opposite,' as Democritus says, to the illuminating body, she receives and admits the sun, so that we should expect her both to appear herself, and to show the sun through. But she is far from doing this; for she is not only invisible herself at those times, but often hides and obscures the sun; 'she cuts off his rays,' as Empedocles says, 'from above towards earth, casting a shadow on so much of the earth as is the breadth of the blue-eyed moon'—just as though the sun's light fell upon night and darkness, not on another star. The statement of Posidonius that it is because of the depth of the moon that the light does not come through her to us is obviously wrong. For the air which is unlimited, and has a depth many times that of the moon, is lit up throughout by the sun and by its rays shining upon it. We are therefore left with the view of Empedocles that the illumination which comes to us from her is caused by a sort of reflection of the sun upon the moon. Hence it is that no heat or brilliance reaches us, as we should have expected if there had been a kindling and mixture of lights. But, just as the echo sent back when voices are reflected is weaker than the original sound, and the blows struck by ricochetting missiles fall with less force, 'so the beam, striking on the broad circle of the moon,' sends to us a feeble and faint stream of light because its force is dissipated owing to the reflection."

21. "... We men are far from thinking the moon, a celestial earth, to be a body without soul and intelligence, and without part in the things of which it is meet to offer the first-fruits to the gods, thereby making, as usage prescribes, a return for the blessings they give us, and, in accordance with our natural instinct, reverencing that which is superior in virtue and power and more honourable. So let us not suppose that we do wrong in making the moon an earth. And, as to this apparent face in her, let us suppose that, just as our earth has certain great depressions, so she is opened up by great depths and clefts containing water or dark air, which the light of the sun does not penetrate or touch, but is there eclipsed, so that the reflection sent hither is scattered."

22. Here Apollonides interrupted, and said: "In the name of the moon herself, pray do you then think it possible that there are shadows of clefts or ravines, and that from thence they reach our eyes, or do you not realize the consequences, and must I tell you of them? Listen, then, though you probably know of them already. The apparent size of the moon's diameter is twelve dactyli (finger-breadths) [$=1°$] at her mean distance. And each of the black and shadowy spots seems larger than half a finger-breadth, so that it must be larger than $\frac{1}{24}$th of the diameter. Again, if we suppose the circumference of the moon to be only 30000 stades, and the diameter therefore 10000, then, on that assumption, each of the shadowy parts would measure 500 stades.

Now consider, first, whether it would be possible for the moon to have depths and unevennesses such as to throw such enormous shadows, and next, if the irregularities are of such size, why do we not see them?" And I, with a smile at him, said: "Well done, Apollonides, to have discovered such a demonstration; for by the same argument you can prove that you and I are taller than the famous Aloades of old, not, it is true, at all times of the day, but rather, for choice, in the early morning or late afternoon—that is, if you really think that, when the sun makes our shadows enormous, this fact furnishes to our sense the splendid inference that, if its shadow is great, the thing which throws the shadow must be enormous. I know that neither of us has ever been at Lemnos, but we have probably both heard the hackneyed iambic line:

'Athos shall hide the flanks of the Lemnian cow':

for the shadow of the mountain falls, it seems, on a certain bronze figure of a heifer, and stretches over the sea for a distance of not less than 800 stades. Do you think that the height which casts the shadow is the cause? And do you forget that it is the distance of the light which makes the shadows of bodies many times longer? Come now, consider the sun at its greatest distance from the moon, when it is full and shows the form of the face most distinctly because of the depth of the shadow. It is the distance of the light itself which has made the shadow great, not the size of the irregularities on the moon's surface. Further, in the daytime the brightness of the sun's rays does not allow the tops of mountains to be seen, but the deep, hollow, and shady places are seen from afar.

It is, therefore, not strange that it is impossible to see quite clearly the manner in which the moon receives the light and is illuminated thereby, whereas the juxtaposition of the shadowy and the bright portions enables us to distinguish them by contrast."

APPENDIX

THE CONSTELLATIONS
OF THE NORTHERN HEMISPHERE

HEMISPHAERIUM BOREALE

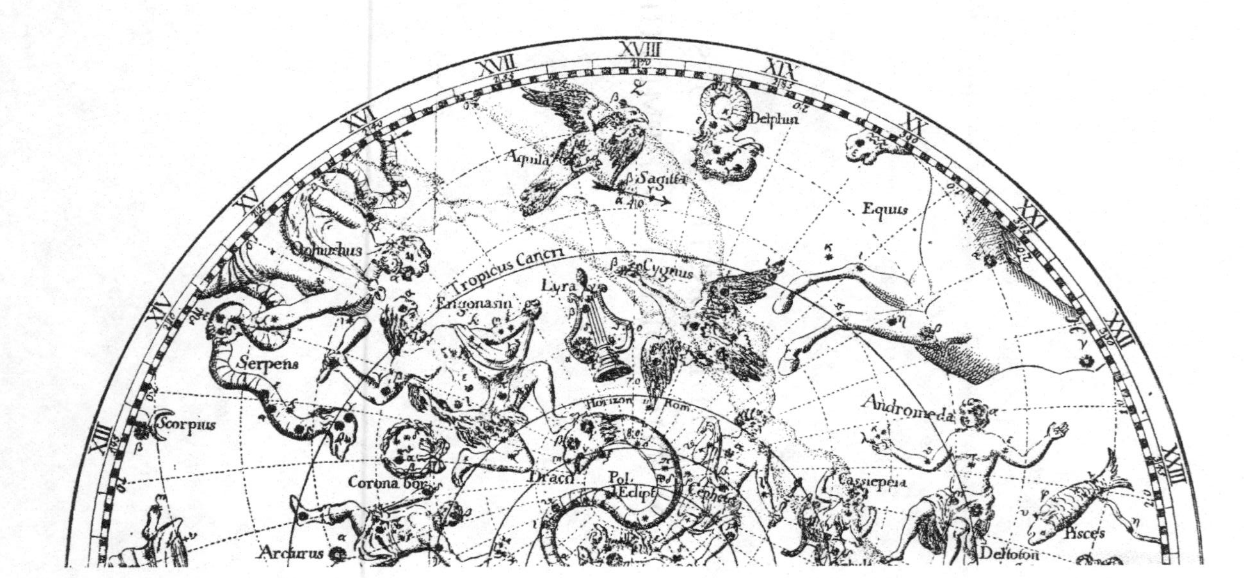

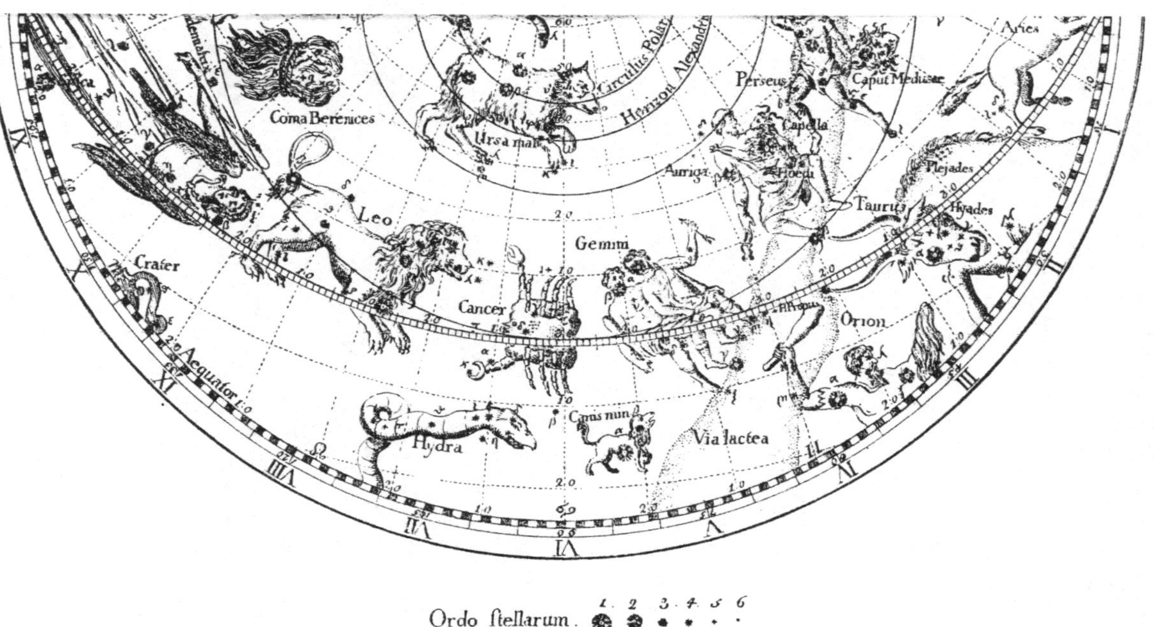

Ordo stellarum. 1. 2 3. 4 5 6

INDEX

K

L

M

N

O

P

MADE AT THE TEMPLE PRESS LETCHWORTH IN GREAT BRITAIN

Printed in the United States
By Bookmasters